Science Achievement in Seventeen Countries

A Preliminary Report

Other Pergamon titles of related interest

GORMAN *et al:* The IEA Study of Written Composition I: The International Writing Tasks and Scoring Scales

Science Achievement in Seventeen Countries

A Preliminary Report

by

International Association for the Evaluation of Educational Achievement (IEA)

PERGAMON PRESS

OXFORD · NEW YORK · BEIJING · FRANKFURT
SÃO PAULO · SYDNEY · TOKYO · TORONTO

U.K.	Pergamon Press, Headington Hill Hall, Oxford OX3 0BW, England
U.S.A.	Pergamon Press, Maxwell House, Fairview Park, Elmsford, New York 10523. U.S.A.
PEOPLE'S REPUBLIC OF CHINA	Pergamon Press, Room 4037, Qianmen Hotel, Beijing, People's Republic of China
FEDERAL REPUBLIC OF GERMANY	Pergamon Press, Hammerweg 6, D-6242 Kronberg, Federal Republic of Germany
BRAZIL	Pergamon Editora, Rua Eça de Queiros, 346, CEP 04011, Paraiso, São Paulo, Brazil
AUSTRALIA	Pergamon Press Australia, P.O. Box 544, Potts Point, N.S.W. 2011, Australia
JAPAN	Pergamon Press, 8th Floor, Matsuoka Central Building, 1-7-1 Nishishinjuku, Shinjuku-ku, Tokyo 160, Japan
CANADA	Pergamon Press Canada, Suite No. 271, 253 College Street, Toronto, Ontario, Canada M5T 1R5

First edition 1988

British Library Cataloguing in Publication Data
Science achievement in seventeen countries.
1. School. Curriculum subjects. Science. Academic achievement of students.
I. International Association for the Evaluation of Educational Achievements
507′.1
ISBN 0-08-036563-9

Printed in Great Britain by A. Wheaton & Co. Ltd, Exeter

CONTENTS

Appendices

Figures

Tables

PREFACE

The International Association for the Evaluation of Educational Achievement (IEA) conducted a study of achievement in science in 19 countries in 1970. This study was repeated in the mid-1980s with 24 countries or systems of education participating. Ten countries took part in both studies.

The IEA is an international association of research centers.IEA does not select countries to participate in its studies. Research centers themselves decide whether or not to participate. A research center from any country may join an IEA project on condition that it has proven experience in doing the sort of work involved in the project. It is the participant members of a project who co-operatively plan the study, the overall framework having been given by the IEA General Assembly - the body comprising all member institutes of IEA. Thus, for example, great care was taken to ensure that the test instruments were equally 'fair' to all countries. It was never a question of a single country's curriculum being used to produce a test and then all other countries having to be tested according to that country's curriculum. This could involve a great deal of test bias and any comparisons between countries' achievement based on that test would be unfair. Similarly, care was taken with all background questions and attitude scales and other measures to ensure that reasonable comparisons could be made.

The work for the Second International Science Study (SISS) was begun in 1980. Like other IEA studies, each of the participating systems collaborated in the research, and each was responsible for its own funding, data collection, data analysis and preparation of national reports. The International Coordinating Center is at the Australian Council for Educational

Research in Melbourne with Dr. Malcolm J. Rosier as International Coordinator. Subsequent data processing and analyses are being undertaken at the Institute of International Education at the University of Stockholm and at the Department of Education of the University of Hamburg.

IEA would like to thank, above all, the National Research Co-ordinators at the national research centers for their commitment to the study and for the painstaking work they have undertaken. Norbert Sellin at the University of Stockholm and his team at the University of Hamburg have achieved a near miracle in the work they have accomplished in two months in checking the data and preparing the summary statistics for use in this report. The data checking involved several stages: checking that the data tapes received from the various countries could be read, ensuring that the data had been arranged according to the international codebook instructions, checking that there were no cases with data outside the valid range for each variable, and making adjustments to the datasets where any problems were encountered. This has literally been a day and night task.

Thanks are also due to Heather Payne and other staff at the Australian Council for Educational Research, and to the Swinburne Institute of Technology in Hawthorn, Australia for providing computing facilities at an earlier stage of the data-cleaning process.

Malcolm Rosier, the international co-ordinator and sampling referee of the study, deserves special praise. His careful work and commitment from the inception of the project have ensured that, despite the severe financial constraints at the international level, the study has progressed satisfactorily through all its steps.

John Keeves, when he was Director of the Australian Council for Educational Research, proposed to IEA that the First Science Study should be repeated. He has had overall responsibility for the study. IEA thanks him for his dedicated work throughout the study and also thanks him for the contribution he has made in many ways to this preliminary report.

The international costs of the study have been supported by the

Australian Council for Educational Research, the Japanese Shipbuilding Industry Foundation, the National Science Foundation (USA), the Bank of Sweden Tercentenary Fund, the Swedish National Board of Education, the Wenner-Gren Foundation, the World Bank, and the generous contributions of several of the research institutions participating in the study in their hosting of meetings. To all of them IEA is grateful.

Finally, thanks are due to Neville Postlethwaite of the University of Hamburg, for his invaluable contribution to IEA and to this science study. He has taken the responsibility for the preparation of this report. It is only through his dedicated work and loyalty to IEA that this report could be published.

This report is an interim presentation of some selected international achievement score results. It is intended to give readers, and in particular educational policy makers, some of the initial information from cross-national comparisons so that questions may be raised concerning future international and national analyses of the data. Further comprehensive reports will be published in 1989. They will be concerned with science education and curriculum in the 24 countries; basic descriptive statistics and explanatory analyses of between and within country differences in science education achievement in the 24 countries; and, a comparison of science education and achievement in the ten countries which were involved in the IEA science studies in 1970 and in the mid-1980s.

Alan C. Purves
Chairman of IEA.
University of Albany
State University of New York

Executive Summary

In the period 1983 to 1986, the International Association for the Evaluation of Educational Achievement (IEA) undertook a study of science achievement in twenty-four countries at three levels in each school system:

1) the 10 year old level - typically Grade 4 or 5;

2) the 14 year old level - typically Grade 8 or 9; and,

3) the final year of secondary school - typically Grade 12.

This preliminary report presents some initial results from seventeen countries. These countries are: Australia, Canada (Eng), England, Finland, Hong Kong, Hungary, Italy, Japan, Korea, Netherlands, Norway, Philippines, Poland, Singapore, Sweden, Thailand, and the U.S.A.

Science achievement tests were constructed collectively by the researchers from the participating countries.

The publication presents the following major results:

1. The validity of the tests.

2. The mean achievement score for each country at each level.

3. For the 14 year olds an extra result of the "bottom 25 percent" of children in school is presented.

4. The achievement differences between boys and girls at each level.

5. The between school differences at each level.

6. The percentage of schools in each country scoring below the lowest school in the highest scoring country.

The following constitutes a selection of findings from the report.

1

2 Science Achievement in Seventeen Countries

1. Ten year olds

The ten year olds in England, Hong Kong, Poland and Singapore have the lowest levels of science achievement and Japan, Korea, Finland and Sweden the highest.

2. Fourteen year olds

At the Grade 8 or 9 level the highest scoring countries are Hungary and Japan and the lowest scoring countries are England, Hong Kong, Italy, the Philippines, Singapore, Thailand and the U.S.A. The U.S.A. was third to last out of seventeen countries with Hong Kong and the Philippines being in the sixteenth and seventeenth places. Thailand had a score equal to that of the U.S.A. but Thailand has only 32 percent of an age group in school, whereas the U.S.A. has 100 percent of an age group in school. The IEA conducted its last survey of science achievement in 1970 and Australia has dropped from third position in 1970 to below the international mean in 1983. The United States has dropped from seventh out of seventeen countries to third from bottom.

Grade 9 in Norway scores lower than Grade 8 in Sweden and Finland although all three countries have the same age of entry to school.

3. Final year of secondary school results

Students were tested in biology, chemistry and physics. In general, Hong Kong, England and Singapore were the highest scoring countries. They also specialize in the sense of studying a very limited number of subjects. Canada (English), Italy, Finland and the U.S.A. are, in general, the lowest scoring countries.

Singapore is the highest scoring country in biology, Japan's scores are remarkably low in biology compared with its performance in physics and chemistry.

In chemistry and physics, Hong Kong is the highest scoring country. Indeed, in physics, Hong Kong Form 6, the penultimate year in secondary school, has a higher score than the final grade in school in all of the other countries at this level. The following table gives the rank order of countries for each population level.

Rank order of countries for achievement at each level

	10 yr. olds Grade 4/5	14 yr. olds Grade 8/9	Grade 12/13 Science Students Biology	Grade 12/13 Science Students Chemistry	Grade 12/13 Science Students Physics	Non-Science students
Australia	9	10	9	6	8	4
Canada (Eng)	6	4	11	12	11	8
England	12	11	2	2	2	2
Finland	3	5	7	13	12	-
Hong Kong	13	16	5	1	1	-
Hungary	5	1	3	5	3	1
Italy	7	11	12	10	13	7
Japan	1	2	10	4	4	3
Korea	1	7	-	-	-	-
Netherlands	-	3	-	-	-	-
Norway	10	9	6	8	6	5
Philippines	15	17	-	-	-	-
Poland	11	7	4	7	7	-
Singapore	13	14	1	3	5	6
Sweden	4	6	8	9	10	-
Thailand	-	14	-	-	-	-
U.S.A.	8	14	13	11	9	-
total no. of countries	15	17	13	13	13	8

4   Science Achievement in Seventeen Countries

4.   Bottom 25 percent of fourteen year olds

The bottom 25 percent of pupils performed particularly badly in England, Hong Kong, Italy, Singapore, and the U.S.A.. The lowest scoring children were scoring at chance level, indicating that from the test's point of view, they were scientifically illiterate.

5.   Sex differences

Boys score higher than girls at all levels. The difference widens from the 10 year old to the 14 year old level (from 0.23 to 0.34 of a standard deviation). At the Grade 12 level the sex differences are 0.17 for biology, 0.36 for chemistry and 0.39 for physics. In biology however, in Sweden and Hong Kong, girls scored higher than boys. Since 1970, there would appear to be little change in the superiority of boys over girls at the 14 year old level and Grade 12 level.

6.   Differences among schools

The differences among schools as a percentage of differences among students varies widely. In Sweden Grade 4 and Japan Grade 5 this percentage is 3 percent and 4 percent respectively. This means that all schools are much the same and that it doesn't make much difference which school a student attends. In the Philippines and Singapore, on the other hand, it makes a lot of difference to which school a pupil goes (56 percent and 39 percent respectively).

At Grade 8/9 level, Norway, Japan, Finland and Sweden have achieved this form of equality of opportunity (2,4,5 and 8 percent respectively), but Singapore (56), the Netherlands (50), the Philippines (48) and Italy (39) have very high values. At Grade 12 level, Hungary, Italy, Japan , Poland and the United States have high differences among schools.

7 .   Low scoring schools

The lowest school mean in the highest scoring country was taken and then the percentage of schools in other countries scoring below that mean was calculated. The table below presents the results.

<u>Percentage of schools scoring lower than the lowest school</u>
<u>in the highest scoring country</u>

	10 yr. olds	14 yr. olds	Biology	Chemistry	Physics
Lowest mean in highest scoring country	12.6	14.8	56.6	30.0	50.5
	Japan	Hungary	Singapore	Hong Kong Singapore England	Hong Kong
Australia	37	8	93	4	67
Canada (Eng)	25	6	95	26	93
England	61	19	14	0	18
Finland	7	2	91	24	93
Hong Kong	77	26	47	0	0
Hungary	21	0	37	13	45
Italy	38	37	100	33	99
Japan	0	1	84	14	36
Korea	7	5	-	-	-
Netherlands	-	16	-	-	-
Norway	58	1	56	4	44
Philippines	83	87	-	-	-
Poland	66	14	48	13	53
Singapore	75	32	0	0	25
Sweden	3	1	78	11	82
Thailand	-	26	-	-	-
U.S.A.	38	30	98	48	89

<u>Comments on achievement in science in different countries</u>

<u>Australia</u>: This country has mediocre results at all three population levels especially in terms of the performance levels of its major trading partners. Its ranking, relative to other countries, has dropped significantly since the first science study in 1970. This warrants further analyses.

6   Science Achievement in Seventeen Countries

Canada (English): Its performance was reasonably good at Grade 9 level but is mediocre at the terminal secondary level.

England: Performance is poor at the upper primary level. Indeed, 60 percent of schools score lower than the lowest scoring school in Japan. At middle secondary school level its performance is still relatively weak. However, in biology, chemistry, and physics at second year sixth level students enrolled in science courses (5 percent of an age group) scored very well compared with other countries. Is it that this system of education is fostering high performance of its elite at the expense of mass education?

Finland: Finland performs very well at Grades 4 and 8, but very poorly at Grade 12, especially in chemistry and physics. The general education priorities set in Finnish education, while well attained in the Comprehensive School, stop short of providing quality training in science at the pre-university level. It is of course a problem, why this is so. The Finnish Upper Secondary School has a strong humanistic and language orientation; it also has a large majority of girls who tend to have lower achievement in science. Furthermore, it places emphasis on biology, geography, and book-oriented methods of teaching, with little place for practical and investigative work. However, since the first science study in 1970, the Finnish Upper Secondary School is now educating double the proportion of the earlier age group.

Hong Kong: Like England, it has low achievement in primary and middle secondary school but performs extremely well, especially in chemistry and physics, at the end of secondary school. Again, like England, it has a high degree of specialization where typically, only three subjects are studied.

Hungary: Its science achievement is high at all levels in the school system and the differences between the sexes are low. Nevertheless, the differences among schools at Grade 12 level are very high and must be an issue for further analyses.

Italy: The relative achievement standing deteriorates from Grade 5 to 12. At Grade 12 level all schools score lower than the lowest scoring schools in Singapore and Hong Kong in biology and physics respectively. The differences among schools are very high at both Grade 8/9 and Grade 12 levels. Much would appear to be amiss in the realm of science education.

Japan: Japan's achievement is very high at Grades 5 and 9 but at Grade 12 it is tenth out of thirteen countries in biology, although only fourth in chemistry and physics. The differences among its schools at Grades 5 and 9 are very low, indicating a high degree of equality of educational opportunity. The differences between the sexes are also very small. However, at Grade 12 level the differences among schools are very high.

Korea: At Grade 5 level, Korea has the highest score together with Japan. At Grade 9 level, it is seventh out of seventeen countries. The differences in achievement between the sexes are quite high but the differences among schools reasonably low. The performance at Grade 12 will be published in the next report of this study.

The Netherlands: This country only tested at Grade 9 level. The grade had an average age of 15 years and 6 months and a very high spread of age. One quarter of this grade sample is aged over 16 years. The differences among schools are very high, indicating an inequality of educational opportunity. This would appear to be an anomaly in educational provision in a country professing equality.

Norway: This country's performance was average compared with other countries. It deliberately tested Grade 9 instead of Grade 8 at the middle secondary school level. Its Grade 9 achievement was inferior to the Grade 8 achievement in both Finland and Sweden. However, both Finland and Sweden have more science lessons in Grades 7, 8, and 9, than does Norway.

8   Science Achievement in Seventeen Countries

The Philippines: The Philippines has very low achievement levels at both Grade 5 and Grade 9. The differences between schools are very high as is the case in many developing countries. There is clearly much work to be done to increase science achievement and to decrease the inequality among schools. It should, however, be remembered that the language of instruction in school is English, whereas the native language is Pilipino.

Poland: Poland improves its achievement relative to other countries from Grade 4 to the final year of secondary school where it performs particularly well in biology.

Singapore: Like England and Hong Kong, Singapore performs relatively poorly at Grades 5 and 9. However, at Grade 13, the small percentage of an age group studying science, typically as one of three subjects, achieves very well and particularly so in biology. There are very large differences among schools at each level which, in a small country, would imply a deliberate policy of differentiation among schools in terms of resources and curriculum.

Sweden: Sweden achieves relatively well at Grades 4 and 8 but not so well at Grade 12. Grade 12 students tend to study many subjects and there would appear to be a deliberate policy of minimizing specialization. Whether this has any effect on the scientific achievement of the top band of scientists in leading industrial positions is something which the Swedes themselves must determine. However, the sex differences in achievement are low and the between school differences are low at Grades 4 and 8.

Thailand: Results from Thailand are only presented for Grade 9 in this report. Its achievement is relatively low but comparable with that of the U.S.A. However, it has only 32 percent of an age group in school at this level.

U.S.A.: The United States of America tries to retain young people in elementary and secondary school. Representative samples of these young people have a middle position in science achievement in Grade 5 and a lower position in Grade 9. The achievement of advanced science students in biology, chemistry, and physics is low. The biology results are especially low. For a technologically advanced country, it would appear that a reexamination of how science is presented and studied is required.

A further general finding is that in the final year of secondary school there was no relationship between the average age of the population (ranging from 17 years and 3 months in Australia to just over 19 years in Hong Kong and Italy) and achievement. Why then do some systems have more years of schooling than others and why do some countries start schooling at an earlier or later age than others?

Further work

The results presented in the booklet show differences between and within countries in science achievement. It is hoped that funding will be forthcoming to undertake further analyses to identify the variables that account for such differences. The explanation of differences will, then, form the content of reports to be published in 1989.

CHAPTER 1: INTRODUCTION TO THE STUDY

This preliminary report presents initial findings from the second international study of science achievement which was conducted by the International Association for the Evaluation of Educational Achievement (I.E.A.) between 1983 and 1986. Achievement results are presented for three school population levels in Chapters 2, 3 and 4. Chapter 5 presents some special analyses on growth in achievement between the population levels, on sex differences in science achievement, on the relationship between measures of the intended curriculum in science and student achievement, and finally, the relationship between the percentage of an age group studying science and achievement.

Further reports of results will appear in 1989. In addition to the presentation of more detailed achievement results, they will discuss factors associated with between country differences and between student differences within countries as well as differences in achievement between 1970 and the mid 1980s. Thus, these volumes will attempt to answer more profound questions about achievement results with the purpose of improving science education.

Twenty-four systems of education participated in the study. Only the data for seventeen countries are presented in this preliminary report. The rest will be included in the 1989 publication; they are not included in this preliminary report because, at the time of going to press, the data files were either not available or not cleaned.

Populations and Samples

Three levels had been selected for testing within each educational system in IEA's first study of science achievement in 1970. The first level was to be near the

11

end of primary school; the second level was to be at a point in secondary schooling where students were still in full-time compulsory education in most countries; and, the third level was the final year of secondary school. This is sometimes known as the pre-university year or the terminal grade of secondary school.

The Second Science Study decided to use the same levels as in the first study so that comparisons could be made, where appropriate, between the first and second studies. The definitions used for the second study were as follows:

Population 1 consists of either all 10 year old students or all students in the grade in which most 10 year olds are.enrolled. In general, this is a point in the educational system where children are still in primary school. The grades are typically Grades 4 or 5.

Population 2 consists of either all 14 year olds or all students in the grade in which most 14 year olds are enrolled. This is a grade level which, in most countries, is near to the end of full-time compulsory schooling. The grades are typically Grades 8 or 9.

The developing countries participating in the study were permitted to test at higher grade levels for Populations 1 and 2 if they considered that the science tests were more appropriate for students at these levels than at Grades 5 and 9 respectively.

Population 3 consists of all students studying science in the final year of secondary school. This is typically Grade 12 or Grade 13. The students in Population 3 were subdivided into three groups:

3B - all students studying biology for examination purposes;

3C - all students studying chemistry for examination purposes;

3P - all students studying physics for examination purposes.

There was also an extra Population 3N consisting of all students not studying any science subject at this level. Not all countries opted to test Population 3N.

Appendix 1 presents a detailed definition of each of the target populations.

Tables 1a and 1b present an overview of the educational systems and populations for which results are given later in this booklet. Table 1a presents

Table 1a: Percentage of an age group in school and mean ages of Populations 1 and 2

	Population 1				Population 2			
	Age of Entry	Grade tested	% in School	Age Mean	Age SDa	Grade tested	% in School	Age Mean	Age SDa
Australia	6	4,5,6	99	10:6	3.3	8,9,10	98	14:5	3.3
Canada (Eng)	6	5	99	11:1	7.1	9	99	15:0	6.1
England	5	5	99	10:3	3.6	9	98	14:2	3.6
Finland	7	4	99	10:10	4.1	8	99	14:10	4.1
Hong Kong	6	4	99	10:5	9.8	8	99	14:7	10.9
Hungary	6	4	99	10:3	5.2	8	98	14:3	4.7
Italy	6	5	99	10:9	5.2	8,9	99	14:7	5.4
Japan	6	5	99	10:7	3.5	9	99	14:7	3.5
Korea	6	5	99	11:2	7.4	9	99	15:0	7.2
Netherlands	6	-	-	-	-	9	99	15:6	12.5
Norway	7	4	99	10:11	4.0	9	99	15:10	4.0
Philippines	7	5	97	11:1	11.3	9	60	16:1	18.9
Poland	7	4	99	10:11	5.4	8	91	15:0	5.8
Singapore	5	5	99	10:10	5.7	9	91	15:3	9.0
Sweden	7	4	99	10:10	4.1	8	99	14:9	3.8
Thailand	6	-	-	-	-	9	32	15:4	8.9
U.S.A.	6	5	99	11:3	6.9	9	99	15:4	9.1

-    = this population not tested

a   = the mean age is presented in years and months and the standard deviation in months

Table 1b: Percentage of an age group in school and mean ages in Population 3

Country	Grade tested	ALL % in School	ALL Agea Mean	ALL SD	3B % in School	3B Agea Mean	3C % in School	3C Agea Mean	3P % in School	3P Agea Mean	3N % in School	3N Agea Mean
Australia	12	39	17:3	11	18	17:1	12	17:3	11	17:3	10	17:6
Canada (Eng)	12/13d	71	18:3	11	28	18:2	25	18:4	19	18:4	-	18:7
England	13	20	18:0	6	4	18:0	5	18:0	6	18:0	10	18:0
Finland	12	45(63)b	18:7	7	45	18:7	14	18:6	14	18:6	-	-
Hong Kongc	12	20	18:4	13	7	18:4	14	18:4	14	18:4	-	-
Hungary	12	18(40)b	18:0	4	3	18:0	1	18:1	4	18:1	9	18:1
Italy	12	52	19:0	13	14	19:5	2	19:3	19	19:3	25	19:1
Japan	12	63	18:2	4	12	18:1	16	18:2	11	18:2	35	18:2
Norway	12	40	18:11	7	10	18:11	15	18:11	24	18:11	-	18:11
Poland	12	28	18:7	5	9	18:8	9	18:7	9	18:7	-	18:7
Singapore	13	17	18:1	8	3	18:0	5	18:0	7	18:0	8	18:3
Sweden	12/13	15(30)b	19:0	11	15	19:0	15	18:11	15	18:11	-	18:11
U.S.A.	12	90	17:7	9	6	17:5	1	17:8	1	17:8	66e	17.2

a   As in Table 1a, the mean age is given in years and months but the standard deviation is in months.

b   In Finland, it is estimated that 63 percent of an age group is in full-time schooling. The vocational schools (18 percent of an age group) were not sampled. In Hungary, 40 percent of an age group is in full-time schooling in the final year of school. However, the vocational schools (22 percent of the equivalent age group) were not sampled. The 18 percent of the age group is those in academic secondary schools studying science. In Sweden, 90 percent of an age group is enrolled in upper secondary education and 80 percent of an age group completes upper secondary education. Fifteen percent are enrolled in science tracks and 15 percent in non-science tracks in Grade 12. The remainder takes a two year vocational or general track and leave school after Grade 11.

c   Hong Kong tested Grades 12 and 13 (Forms 6 and 7). The average age for Grade 12 is ten months less than at Grade 13. The percentage studying the science subjects is slightly more than that at Grade 13.

d   In Ontario, it was Grade 13.

e   In addition, there are students taking first year physics courses.

-   this population not tested (except for 3N in Canada (English) and Norway).

information on Populations 1 and 2, while Table 1b presents information on Population 3. All of the information in Tables 1a and 1b was supplied by the national research centers involved in the study. Some comments should be made about the target populations considered in these two tables.

The Netherlands only tested at Population 2 level. The Philippines did not test Population 3 and Thailand did not test Population 1. The percentages of an age group in the different Population 3 groups do not necessarily add up to the proportion of an age group in school because some students take two or more science subjects and are, therefore, counted twice. For Populations 1 and 2, Australia had strict age samples whereas other countries decided to test the modal grade of 10 year olds and 14 year olds, that is, the grade in which most 10 or 14 year olds were to be found.

It will be noted that the age of entry to school is different and ranges from 5 to 7 years old.

At Population 1 the Philippines tested Grade 5 instead of Grade 4 because the international tests were more appropriate for Grade 5. It should also be pointed out that the tests were administered to bilingual students. In the Philippines, English is the medium of instruction but Pilipino is the native language. Canada (Eng), Korea, and the U.S.A. had mean ages of just over 11 years at the date of testing. However, the students had been ten years of age throughout most of the grade.

For Populations 1 and 2, the percentages of an age group in school at the grade level tested was set at 99 percent (indicating that there is only a small number of students not in school) unless a system gave precise figures.

At Population 2 level, Norway and the Philippines have high mean ages. Norway deliberately tested Grade 9 instead of Grade 8 since Grade 9 is the last grade of basic schooling. The Netherlands has an average age of 15 years and 6 months. Twenty-five percent of the Netherlands grade group is 16 years or older. This is probably due to grade-repeating and foreign students who are placed in lower grades in order to keep up with the program.

For Population 3 - the terminal secondary grade - the percentage of an age group in school varies from system to system and ranges from 17 percent in

16 Science Achievement in Seventeen Countries

Singapore to 90 percent in the U.S.A.. The U.S.A. figure denotes the percentage of an age group enrolled in Grade 12 and not the percentage graduating from that grade. The percentages of an age group and the mean ages of those studying science (biology, chemistry, and physics) are presented as well as for those in the group of students not studying science. In some countries where no national statistics were available for the numbers of science students and 3N students, estimates were made by the national centers to arrive at the proportion of an age group in the different groups.

The average age of students in the last grade in school varies considerably. The youngest group of students is in Australia (17 years and 3 months). The United States twelfth graders are also only 17 years and 7 months old. In all Population 3 tables two sets of results are presented for Hong Kong. One is for Form 6 secondary (Grade 12) and its average age is nearly 18 years and 4 months. The second set of results is for Form 7 (Grade 13) which is ten months older. In Hong Kong there are two pre-university curricula: a one year Form 6 course leading to an entrance qualification for the Chinese University of Hong Kong, and a more dominant two year Forms 6 and 7 course leading to an advanced level examination which can serve as an entrance qualification to all tertiary institutions.

Appendices 2 to 7 present the number of schools and students tested at each population level and the appropriate response rates. All of the samples were probability samples. In most cases a probability sample of schools was drawn first and then either a whole class or up to 24 students were drawn at random within the schools. In two or three countries a first stage of sampling consisted of drawing a sample of provinces or districts or counties. Then, within selected provinces, districts or counties, schools were drawn at random. In general, the samples mirror the target populations well. The standard errors of sampling of a population mean science score for all populations tested are given in Appendix 9. A separate technical report (Rosier, in preparation[1]) presents details of the sampling and the

---

[1] Rosier, M.J.: Sampling and Administration for the Second   International Science Study (in
                   preparation)

administration of the study.

Appendices 2 to 7 also presents the school and student response rates for Populations 1 and 2 and the various sub-populations in Population 3. To be accepted into the final data set (achieved sample), a school had to have a minimum of three students included in the final data set. In addition, a student had to have attempted at least three items on the common test to be included on the data files. The response rates for both schools and students are presented in the form:

$$\frac{\text{achieved sample}}{\text{designed sample}} \times 100$$

At an early stage of the study each national center was required to produce a sampling plan and the designed sampling figures used for the calculation of the response rates are those given in each nation's sampling plan.

Appendices 2 to 7 have been arranged to show, for the schools and the students in each country's target population, the size of the target population; the number in the designed sample, the executed sample (that is the number of schools or students recorded on the data file sent in by countries), the achieved sample and finally the response rate. Population 3 is a little complicated in that in some cases the target population figures were known and designed samples could be created, but in other cases the actual number of students studying biology, chemistry, or physics or no science at all was not known. Indeed, given the various combinations of subjects being studied in the final year of secondary school, it is, in general, a matter of drawing a probability sample of schools and then testing those students in each of the science groups. In the percentages of an age group given in Table 1b it was pointed out that there was double counting in that the same student could be included in the physics group and chemistry group. Appendices 4 to 7 present the sample sizes for each group separately. Singapore, it should be pointed out, tested everyone in the last grade of school.

Finally, it should be noted that the proposed number of students for the designed sample of students was typically an estimate since it was not known ahead of sampling design exactly how many students in each school would enter the sample. Nevertheless, the estimates for each country are expected to be very near the actual number.

To give readers an idea of the magnitude of the study, the number of schools and students at each population level from which data were collected in the seventeen countries in this report were:

	Schools	Students
Population 1	2617	71576
Population 2	2819	73001
Population 3	4372	76271
Total	9808	220848

However, it should be noted that, for Population 3, many of the same schools appear in each of the subject samples.

Construction of the Science Tests

Table 2 on page 16 presents 57 content areas that were judged by the IEA Science Committee to represent the major content areas likely to be taught in science at school. This can be conceived of as a common science curriculum across the world. This common curriculum was, however, not dreamt up in an ivory tower by a group of science educators not knowing what goes on in science classes in schools. In the period 1967 to 1969 an international test of science had been constructed. It was based on the analysis of the science actually taught in schools at the 10 year old, 14 year old and terminal grade levels. In 1980-81, further content areas were suggested as likely to have entered the school curricula. When these additional content areas were checked against school science curricula in the different countries, it was discovered that they had only entered with sufficient emphasis into the science curricula in a very few countries. They did not exist in the curriculum of enough countries to warrant their inclusion in the testing program to

Table 2: Content Areas, Curriculum Emphasis, and Number of Items: Populations 1, 2 and 3

Content Areas	Population 1		Population 2		Population 3		Biology		Chemistry		Physics	
	Grid Mean	No.of Items	Grid Mean	No.of Items	Grid Mean	No.of Items	Grid Mean	No.of Items	Grid Mean	No.of Items	Grid Mean	No.of Items
Earth Science												
1 Solar System	1.6	3	1.8	3	1.5	1						
2 Stellar System	0.9	1	1.1	0	1.0	1						
3 Meteorology	2.2	3	1.8	4	1.5	1						
4 Constitution of Earth	1.1	0	1.6	0	1.6	0						
5 Physical Geography	1.3	1	1.6	2	1.6	1						
6 Soil Science	1.2	0	1.4	0	1.2	0						
Total	1.4	8	1.5	9	1.4	4						
Biology												
7 Cell Structure and Function	1.0	0	2.5	0	2.6	0	3.0	0				
8 Transport and Cellar Material	0.1	0	1.4	0	1.7	0	2.8	2				
9 Cell Metabolism	0.3	0	2.3	0	2.4	0	3.0	0				
10 Cell Responses	0.1	0	1.2	1	1.3	1	2.5	0				
11 Concept of the Gene	0.1	0	1.2	0	1.5	0	2.9	6				
12 Diversity of Life	2.2	4	2.4	2	2.3	0	2.9	1				
13 Metabolism of the Organism	1.3	3	2.5	7	2.5	4	3.0	3				
14 Regulation of the Organism	0.1	0	1.6	1	1.7	1	2.9	3				
15 Co-ordination/ Behaviour of the Organism	1.1	0	1.5	1	1.4	0	2.7	3				
16 Reproduction and Development of Plants	2.3	6	2.5	3	2.5	0	2.8	1				
17 Reproduction and Development of Animals	2.3	5	2.0	2	2.4	2	3.0	2				
18 Human Biology	1.9	4	2.4	3	2.5	0	2.9	2				
19 Natural Environment	2.5	0	2.0	2	1.9	0	3.0	1				
20 Cycles in Nature	2.1	0	2.0	1	2.3	1	2.9	0				
21 Natural Groups and their Segregation	0.2	0	0.3	0	0.6	0	2.4	0				
22 Population Genetics	0.1	0	0.4	0	0.8	0	2.5	0				
23 Evolution	0.3	0	1.5	0	1.4	0	2.7	6				
Total	1.0	22	1.6	23	1.9	9	2.8	30				
Chemistry												
24 Introductory Chemistry	1.2	2	2.8	7	2.8	3			3.0	3		
25 Electro-chemistry	0.0	0	1.8	0	2.0	1			3.0	4		
26 Chemical Laws	0.0	0	1.5	2	1.6	2			2.9	3		
27 Chemical Processes	0.1	0	2.2	1	2.3	0			2.4	4		
28 Periodic System	0.1	0	1.6	0	2.0	0			3.0	2		
29 Energy Rel.hips in Chem. Systems	0.0	0	1.2	0	1.2	0			2.6	1		
30 Rate of Reactions	0.1	0	0.9	0	0.9	1			2.5	2		
31 Chemical Equilibrium	0.0	0	0.2	0	0.6	0			2.9	2		
32 Chemistry in Industry	0.6	1	1.2	1	1.5	0			2.3	1		
33 Chemical Structure	0.1	0	2.2	2	1.9	1			2.9	2		
34 Descriptive Inorganic Chemistry	0.4	1	1.8	1	2.1	0			2.8	2		
35 Organic Chemistry	0.7	0	1.2	1	1.3	0			2.8	2		
36 Environmental Chemistry	0.1	0	1.3	0	1.2	0			1.3	0		
37 Chemistry of Life Processes	0.3	1	1.4	0	1.3	1			1.4	0		
38 Nuclear Chemistry	0.0	0	0.1	0	0.5	0			1.6	2		
Total	0.2	5	1.4	15	1.5	9			2.5	30		
Physics												
39 Measurement	2.3	2	2.6	3	2.5	0					2.9	3
40 Time and Movement	1.9	1	2.0	0	2.1	1					3.0	2
41 Forces	1.4	4	2.6	4	2.5	2					2.8	1
42 Dynamics	0.2	1	1.8	0	2.2	0					3.0	4
43a Energy	1.5	0	2.5	0	2.4	0					2.9	3
43b Machines	1.8	2	1.7	2	1.8	0					1.4	0
44 Mechanics of Fluids	0.9	1	1.6	1	1.4	0					1.3	0
45 Introductory Heat	1.8	2	2.5	3	2.4	1					2.3	1
46 Change of State	2.1	1	2.3	2	2.4	0					2.1	1
47 Kinetic Theory	0.7	0	1.7	1	1.6	1					2.5	1
48 Light	1.3	1	2.1	1	1.7	0					2.5	1
49 Vibration and Sound	1.3	2	1.9	2	1.8	0					2.5	1
50 Wave Phenomena	0.1	0	0.5	0	0.9	0					2.9	3
51 Spectra	0.7	0	1.1	0	0.8	0					2.6	1
52 Static Electricity	0.6	1	1.9	0	1.9	0					2.7	1
53 Current Electricity	1.6	1	2.5	3	2.3	2					2.9	2
54 Electromagn./Alternating Currents	1.6	2	2.0	1	1.9	0					3.0	2
55 Electronics	0.1	0	1.1	0	1.0	0					1.8	1
56 Molecular/Nuclear Physics	0.2	0	0.6	0	1.0	1					2.5	2
57 Theoretical Physics	0.0	0	0.1	0	0.2	0					1.5	0
Total	1.1	21	1.3	23	1.7	8					2.5	30

take place in 1983-84. Each content area could be broken down in terms of simple information, the understanding of a principle, and the application of information and understanding to solve a practical problem.

There had been two criticisms of the 1970 international test: 1) some items were too "wordy" so that one wasn't sure whether science or reading skill was being measured; 2) the items that had been included in the paper and pencil test employed to serve as a proxy measure of achievement in practical work in science did not measure validly either practical science or science achievement. These two criticisms resulted in new items being written

It must be emphasized that the tests are international tests. They are not tests based on the curriculum of one educational system with the expectation that other systems use those tests regardless of how much overlap the first system's curriculum has with other systems' curricula. Indeed, the development of all instruments was a collaborative effort involving all educational systems in the study at that time. This is essential for the validity of the cross-national comparisons that are made. A common curriculum grid was developed using the curricula of all systems participating in the study; items were provided by all systems for the measurement of particular cells in the grid; any new item was trialled in at least five different systems. This common approach resulted in a test which was regarded as equally "fair" or "unfair" to all systems. Given that time for testing was restricted in many school systems, a decision was taken to structure the items into a core (or common test) and a set of rotated tests. Thus, for Population 1 there was a core test (taken by all students) consisting of 24 items and four shorter tests of 8 items each. These shorter tests were rotated so that two were taken by each student. For Population 2, there was a core test of 30 items and four rotated tests of ten items each. At Population 3 level there was a core test (designated Test 3M) of 30 items which was taken by all students. For the physics, chemistry, and biology students, there were 30 items in each of their respective separate tests. For the non-science students there was a 3N test of 30 items.

Test Validities

Three indices of test validities were constructed. The first index was a curriculum relevance index which measured the extent to which the test covered the national curriculum. In other words, it is an attempt to examine to what extent the tests are of equal validity for the different countries.

The second index was a test relevance index which measured the extent to which the items in the test were appropriate to the national curriculum. The third index was a curriculum coverage index which measured the extent to which the total national curriculum covered the curriculum of all countries. A detailed account of the construction of the measures and the resultant indices is given in Appendix 8.

It should be emphasized that, especially in countries without any national curriculum, difficulties were experienced in arriving at compromises for the curriculum ratings, so that the overall estimates must be regarded as only a crude measure.

It is the curriculum relevance index which is the most important of the indices for judging the validity of the tests. In the following remarks, it is only the core test curriculum relevance indices that are mentioned. However, Appendix 8 also presents all three indices for the core tests plus the rotated tests.

The Population 1 indices range from 0.28 to 0.51 with most countries clustering between 0.35 and 0.45. Thus, although the test does not cover all of the national curricula (given the 24 item test), the test would seem to have approximately the same validity for the different countries involved. It must be remembered that in some primary schools science is not taught as a specific subject.

The Population 2 indices for the core test range from 0.32 to 0.44 with most countries clustering at around 0.36. The Population 3 indices for biology are virtually identical at 0.65, for chemistry at 0.93, and for physics at 0.89. The Population 2 core test can be said to be equally fair or unfair even though the overall index is not high. The Population 3 tests cover nearly all of the curricula in the separate subjects.

SASC—C

It should also be pointed out that all indices for the core plus rotated tests are higher in all cases than for the core test alone. Moreover, the test relevance (an appropriateness measure) of the test items is very high.

The validity of the core test for national curricula can be said to be more or less equally "fair" or "unfair" for Populations 1 and 2 although the values of the indices are somewhat low. The validities for the Population 3 science students' tests are high and very similar. In Chapter 5 relationships between these indices and student achievement on the core and separate science tests are considered.

Test Reliabilities and Standard Errors of Sampling

Appendix 9 presents the reliability coefficients (KR-20) for the tests used in this report. Given that only the core tests have been used for Populations 1 and 2, it is to be expected that the reliability coefficients will be smaller than for the total test, that is core test plus rotated tests. At Population 3, the separate science tests are the actual tests and not a subsample of a larger battery of tests. The standard errors of sampling of national means were calculated using the jackknife procedure based on the deletion of one primary sampling unit (the school) at a time. It is suggested that readers refer to the standard errors when examining possible differences between country scores.

The Comparisons to be Made

The large probability samples used in this IEA study allow accurate estimates to be made of the national mean scores of each national system of education. Many senior educational policy makers in various countries have requested evidence on how their country achieves when compared with other countries. They are interested in the mean level of performance and the range of performance for particular grade levels or age groups. They also request comparative information on 'how the bottom group performs'.

Chapters 2, 3 and 4 present this kind of information. In each chapter, the results for the students in each target population are presented both in tabular and diagrammatic form. The student data are followed by some information on between school differences in achievement. The performance of students on some selected items from the science tests is presented in Appendix 12.

When making comparisons, it should be borne in mind that the test measures being used contain only 24 items in the case of Population 1 and 30 items in the case of Populations 2 and 3. It should be emphasized that the science measured in the tests is general in that it is based on common elements in all systems and avoids any aspect of science occurring in only one country. Information on the validity of the measures employed for the evaluation of each national curriculum has been given earlier in this chapter and in Appendix 8.

Finally, although each target population was defined internationally, it can be seen in Tables 1a and 1b that there are differences in the mean ages of the national samples. Hence, due caution should be exercised when making comparisons between countries with large age differences.

With these caveats, it is for each reader to determine the meaningfulness and validity of the comparisons made.

CHAPTER 2: POPULATION 1 TEST SCORES

It will be recalled that Population 1 was defined as either 10 year olds or all children in the grade where most 10 year olds were to be found in the system. Population 1 for all countries is either in elementary school or in the primary grades of the comprehensive school system. In some countries, science is not taught as a separate subject but as part of general or social studies.

This chapter presents information on the scores of the students on the 24 item core test followed by information on achievement differences among schools. Information about performance on selected items is presented in Appendix 11.

Student Core Test Results

Table 3 presents the mean score and standard deviation of scores for each country. It also presents the score point where the student at the 25th percentile occurred, the median and the 75th percentile achieved by a student. Finally, the mean age in years and months of the students tested in a country is given.

The Population 1 core test has 24 items. The national mean scores range from 9.5 in the Philippines to 15.4 in Japan and Korea. Indeed, with the exception of the Philippines, the country scores appear to cluster within a standard deviation of each other. However, as shown in Chapter 5 this represents three to four years of schooling. Most of the countries have a mean age between 10:0 and 10:11 but Canada(Eng) and the U.S.A. are

Table 3:   Student Achievement Score Data for Population 1
(24 item Core Test)

Country	N	Mean	SD	25%	Median	75%	Mean Age
Australia	4259	12.9	4.5	10	13	16	10:6
Canada(Eng)	5104	13.7	4.3	11	14	17	11:1
England	3748	11.7	4.5	8	12	15	10:3
Finland	1600	15.3	4.0	13	16	18	10:10
HongKong	5342	11.2	4.2	8	11	14	10:5
Hungary	2590	14.4	4.5	12	15	18	10:3
Italy	5156	13.4	4.7	10	14	17	10:9
Japan	7924	15.4	4.0	13	16	18	10:7
Korea	3489	15.4	4.2	13	16	18	11:2
Norway	1305	12.7	4.1	9	13	15	10:11
Philippines	16851	9.5	4.5	6	9	12	11:10
Poland	4390	11.9	4.5	8	12	15	10:11
Singapore	5547	11.2	4.1	7	10	13	10:10
Sweden	1449	14.7	4.0	12	15	18	10:10
U.S.A.	2822	13.2	4.6	10	13	17	11:3

slightly over 11 years of age whereas the Philippines has an average age of
11 years and 10 months and the students are clearly a year older than the
students in the other countries. In the Philippines, many children do not
enter school at the official age of entry. Some students achieved the
maximum score of 24 in all countries except for England, Norway, and
Singapore where it was 23. The minimum score was 0 in England, Hong
Kong, Hungary, Italy, the Philippines, Poland, and Singapore. In other
countries it was 1 or 2

There is little to choose among the Japanese, Koreans and Finns with
the highest scores. It is of interest that England, Singapore, and Hong Kong
have relatively low scores and one wonders if this might have to do with the
English science curriculum or with the validity of the tests or both. Further
analyses are required to examine this point in more detail.

Figure 1 presents the same results as in Table 3 but in diagrammatic
form. The diagram is known as a box and whisker plot. The length of the box
in the middle represents the range of scores from the 25th to 75th

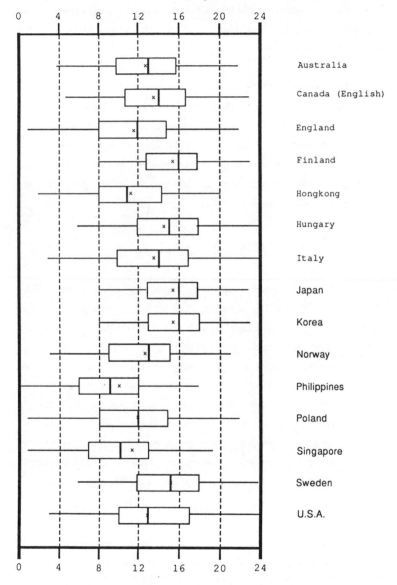

Figure 1: Core Test for Fifteen Countries (Population 1)

percentile. The median is shown as a vertical straight line within the box and the mean score is represented by a small star. The whisker to the left of the box represents the distance from the 25th percentile to that score point located by subtracting the interquartile range (75th percentile point minus the 25th percentile point) from the 25th percentile. The length of the whisker to the right represents the distance from the 75th percentile to the score point located by the 75th percentile plus the interquartile range or the maximum achieved score, whichever is larger. (For further details, see Tukey, 1977[1]). It should be noted that, for convenience, in this diagram the outliers or extreme data points are not given. The comments on Figure 1 are the same as for Table 3.

School   Achievement

Table 4 presents similar data to those given in Table 3 but this time for the means of schools. The mean score which is the same as the student mean score is presented together with the standard deviation of school means. The minimum, median and maximum scores are presented. The next column presents the intra class correlation (roh) (Kish, 1987[2]). This is an indicator of the extent to which students within each country cluster within schools. For example, 0.15 would mean that 15 percent of variance is between schools and 85 percent is between students within schools. The final column gives the percentage of schools in each country scoring lower than 12.6 which is the mean score of the lowest performing school in Japan which had the highest mean score of all countries in Population 1.

---

[1]   Tukey, J.W. (1977) Exploratory Data Analysis.
       Reading, Mass..Addison-Wesley

[2]   Kish, L. (1987) Statistical Design for Research.
       New York. John Wiley.

Table 4: School Achievement Score Data for Population 1
(24 item Core Test)

Country	N	Mean	SD	Min	Median	Max	Roh	Percent of school means below 12.6
Australia	220	12.9	2.0	7.3	13.1	18.9	.15	37
Canada(Eng)	215	13.7	1.7	5.8	13.5	19.4	.12	25
England	181	11.7	2.1	5.6	11.9	17.2	.17	61
Finland	106	15.3	1.5	7.5	15.4	20.0	.07	7
Hong Kong	137	11.2	2.3	6.0	10.9	21.8	–	77
Hungary	100	14.4	2.3	6.3	14.3	20.5	.22	21
Italy	119	13.4	2.2	6.9	13.5	21.8	.19	38
Japan	221	15.4	1.0	12.6	15.4	20.2	.04	0
Korea	146	15.4	1.9	11.0	15.4	21.5	.16	7
Norway	91	12.7	1.9	7.5	12.3	20.0	.15	58
Philippines	463	9.5	3.4	4.2	8.6	21.0	.56	83
Poland	199	11.9	2.3	7.0	11.8	20.7	.22	66
Singapore	232	11.2	2.6	5.1	10.2	18.2	.39	75
Sweden	64	14.7	1.1	11.9	14.7	17.1	.03	3
U.S.A.	123	13.2	1.9	8.1	13.2	17.2	.14	38

The standard deviation of school means ranged from 1.0 in Japan to 3.4 in the Philippines. Another way of looking at this is to examine the roh values. The differences in science achievement between schools as a percentage of the total between student variance ranges from 3 percent in Sweden and 4 percent in Japan (that is very little differences between schools in their achievement) to 56 percent in the Philippines and 39 percent in Singapore. The figure for the Philippines is high but given the big differences between urban and rural schools it was expected to be high but not as high as 50 percent. Singapore's between-school differentiation policy leads to this high roh value. Sweden has a policy of promoting as little difference as possible between schools and the success of this policy can be seen. Japan's success is equally impressive.

The percentage of schools in other countries scoring lower than the lowest scoring school in Japan is given in the last column of Table 4. The large percentages in England, Hong Kong, Poland and Singapore are surprising and will surely be a point for discussion in those countries. The percentages for Sweden, Finland and Korea are low.

Conclusion

Science for 10 year olds in school is taught as a separate subject in some schools and as part of general or social studies in others. Japan, Korea, and Finland have high achievement levels and England, Singapore and Hong Kong have low achievement scores The magnitude of the difference in levels of achievement in these two groups of countries represents between three and four years of schooling. Consequently, the difference may well be related to the extent of formal teaching in science in the schools of the countries in the two groups.

Do those who perform poorly at age 10 catch up in other secondary grades or not? Chapter 3 examines the results for 14 year old students.

CHAPTER 3: POPULATION 2 TEST SCORES

The presentation of the core test scores for Population 2 is made in an identical way to that of Population 1 but this time for seventeen countries. The two extra countries are the Netherlands and Thailand. First, the achievement is presented for the total group. Then, the performance of the bottom 25 percent of the group is examined. This is followed by a special examination of how many learn how much (the "yields" of education). Finally, information is presented on school differences in achievement.

Student Core Test Results

Table 5 presents the achievement data for Population 2.

The mean scores range from 11.5 in the Philippines to 21.7 in Hungary. With the exception of the Philippines, the range of mean scores is just over one standard deviation. Compared with Population 1, the rank order of Poland is higher and that of Korea is lower. England is slightly higher in the rank order but there is very little difference among the scores for England, Italy, Thailand, Singapore, and the U.S.A., all of which can be regarded as having about the same level of performance. Norway has kept the same relative position but it should be remembered that Norway's population is nearly one year older than most of the other countries. Given this fact, it compares very poorly with the two other Scandinavian countries. The standard deviations range from 4.1 in Thailand to 5.1 in the Netherlands. The average of all countries is 17.6, and for all countries minus the two so-called developing countries of the Philippines and Thailand it is 18.1. The

31

"developed" countries, therefore, should be comparing themselves with the overall mean of 18.1.

It should be recalled that in the two developing countries in Population 2 the Philippines has 60 percent of an age group in school and Thailand has only 32 percent in school. All of the other countries have over 90 percent of an age group in school. The maximum score of 30 was achieved in all countries except in England, Italy, the Philippines, and the U.S.A. where the maximum score was 29; in Thailand it was 28. No student scored zero. The lowest score in Finland was 4 and in Korea, Norway, and Poland it was 3. In other countries it was 2 or 1.Again, it must be recalled that the student had to have attempted three items, but not necessarily have given the correct answers, to enter the data set.

It should be mentioned that, between 1970 when the first IEA study took place and the mid 1980s when the present study's data were collected, the ordering of some countries in Table 5 was reversed. Hungary ranked

Table 5:   Student Achievement Score Data for Population 2 (30, item Core Test)

Country	N	Mean	SD	25%	Median	75%	Mean Age
Australia	4917	17.8	4.9	15	18	22	14:5
Canada (Eng)	5543	18.6	4.7	15	19	22	15:0
England	3118	16.7	4.9	13	17	20	14:3
Finland	2546	18.5	4.2	16	19	22	14:10
Hong Kong	4973	16.4	4.5	13	17	20	14:7
Hungary	2515	21.7	4.7	19	22	25	14:3
Italy	3228	16.7	5.0	13	16	20	14:7
Japan	7610	20.2	5.0	17	21	24	14:7
Korea	4522	18.1	4.6	15	18	21	15:0
Netherlands	5025	19.8	5.1	16	19	23	15:6
Norway	1420	17.9	4.7	15	18	21	15:10
Philippines	10874	11.5	4.6	8	11	14	16:1
Poland	4520	18.1	5.2	14	18	22	15:0
Singapore	4430	16.5	4.9	13	16	20	15:3
Sweden	1461	18.4	4.9	15	19	22	14:10
Thailand	3780	16.5	4.1	14	17	19	15:4
U.S.A.	2519	16.5	5.0	13	17	20	15:4

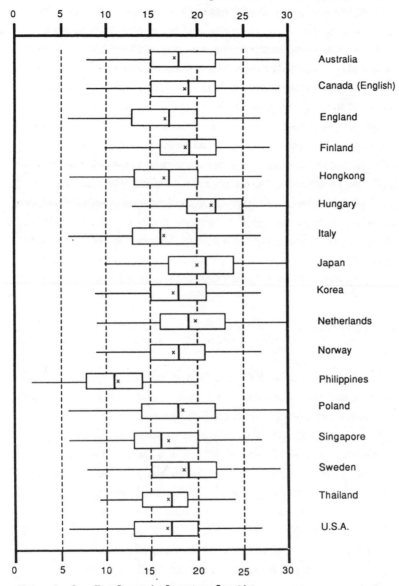

Figure 2: Core Test Scores for Seventeen Countries

34   Science Achievement in Seventeen Countries

second behind Japan in 1970 but was ranked first above Japan in the 1980s. The reasons for this rise, possibly caused by the improved performance of girls in Hungary (see Chapter 5), warrant further detailed analyses. Likewise, it should be noted that whereas Australia held the third rank in 1970, it had fallen to below the middle of the listing in the 1980s. Detailed analyses will be required to determine whether the decline of the Australian students in science is associated with curriculum changes or whether other factors account for this apparent slippage in achievement. Furthermore it might be that there are large differences between the states in Australia and this is a point that will doubtlessly be taken up in the Australian national analyses.

The low level of achievement in the United States should also be noted and detailed analyses of the different data sets (1970, 1983/84, and 1986) should be carried out in a quest for reasons for this low level of achievement.

Figure 2 presents the same data as in Table 5 but in diagrammatic form.

Bottom 25 Percent

The above scores have dealt with all the students in school at the Population 2 level. There has been increasing concern in the 1980s about the lowest achieving groups in compulsory schooling. This might be the bottom 15 percent, or 20 or 25 or 30 percent. In this preliminary report, the bottom 25 percent of those in school is taken and the distributions examined.The twenty-fifth percentile score was taken and the box and whisker plots calculated for all students at the 25th percentile and below.

It is often hypothesized that the group at the bottom of the school is illiterate, innumerate, bored at school, and/or uninterested in learning anything. How, then, does the lowest achieving 25 percent of students in Population 2 score on the science core test? Table 6 presents the data for these low achieving students. Figure 3 presents the same results

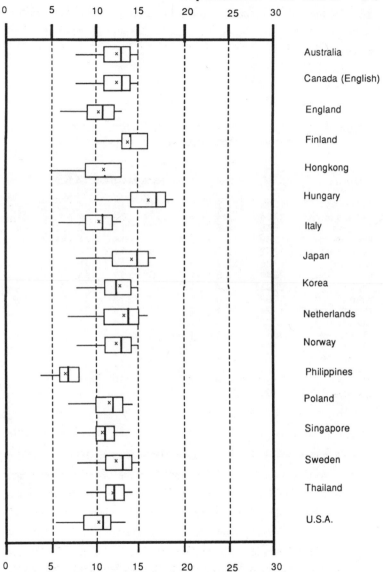

Figure 3: Core Test Science Scores of the Bottom 25 Percent
of Those in School at Population 2 Level

diagrammatically. The standard errors of sampling of the mean values recorded in Table 6 and Figure 3 are twice those given in Appendix 9 for Population 2.

Table 6:   Student Achievement Score Data for Bottom 25 Percent of Population 2 (30 item Core Test)

Country	N	Mean	SD	25%	Median	75%
Australia	1395	12.2	2.6	11	13	14
Canada (Eng)	1484	12.4	2.5	11	13	14
England	845	10.5	2.2	9	11	12
Finland	789	13.7	2.4	13	14	16
Hong Kong	1292	10.7	2.2	9	11	13
Hungary	726	15.7	3.2	14	17	18
Italy	948	10.5	2.2	9	11	12
Japan	2098	13.7	3.0	12	15	16
Korea	1303	12.4	2.4	11	13	14
Netherlands	1524	13.1	2.8	11	14	15
Norway	436	12.5	2.3	11	13	14
Philippines	3092	6.4	1.5	6	7	8
Poland	1144	11.4	2.4	10	12	13
Singapore	1184	10.7	2.0	10	11	12
Sweden	394	12.2	2.5	11	13	14
Thailand	1154	11.8	2.2	11	12	13
U.S.A.	728	10.3	2.5	9	11	12

It is clear that in Hungary, Japan and Finland the lowest achieving twenty five percent of students compare favorably with other countries. However, in England, Hong Kong, Italy, the Philippines and the U.S.A. the bottom quarter of the bottom twenty five percent are not performing well. They are only just scoring above chance level. (On a 30 item test with five options per item there is a one in five chance that students will select the correct answer by guessing. Thus, a score of six could be achieved by guessing only).

Yield Histograms

Figure 4 presents the yield histogram for each country. If a country's system of education is extremely efficient all students would get every question right (a single rectangle) but this is not the case. If a histogram is drawn indicating what percent of students achieve no correct response, one correct response, two correct responses and so on, we gain an impression of how many are achieving how much and the histogram is the 'yield' for achievement in science education at this level in each country's school system. The area of the histogram can be expressed as a percentage of the maximum yield for achievement in science (i.e. for all children answering all items correctly). This percentage has been named an achievement efficiency index. This index is given under each histogram

In several of the yield histograms it will be seen that there are two histograms in one diagram. One is shaded and the other is black. The shaded histogram represents the yield for all the scores of all of the students in school. The black histogram represents the scores of that group when it is calculated as the proportion of an age group (see Table 1a). Thus, in the Philippines the shaded histogram represents all of those in school (and this is set at 99 percent) but in the Philippines only 60 percent of an age group is in school and that yield is presented in black bars. In Canada (Eng) 99 percent of an age group is in school, so no shaded bars have been presented. The two countries where there are large differences are in the Philippines and Thailand. The Philippines, as already mentioned, has only 60 percent of an age group in school and Thailand has only 32 percent in school. Of those in school in Thailand the achievement efficiency index is 56 percent which is the same as the U.S.A. with 56 percent but when calculated as a percentage of an age group it is reduced to 18 percent.

The black bar histogram represents the achievement yield of the proportion of an age group enrolled in school. It does not follow that those not in school know no science, but the index does act as a pointer to what might be achieved if a higher proportion of an age group was enrolled in

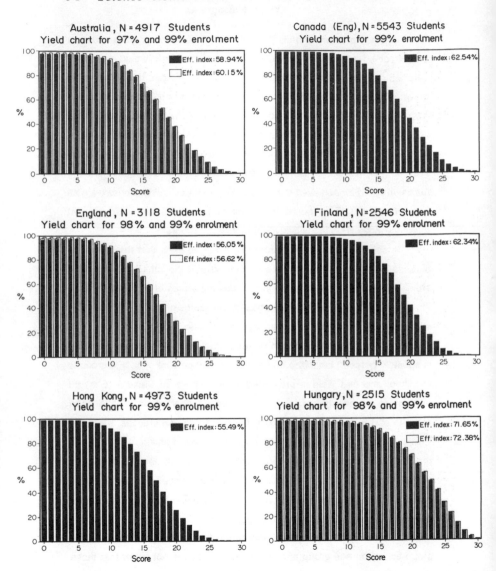

Figure 4: Yield Histograms and Achievement Efficiency Indices for 17 Countries (Population 2)

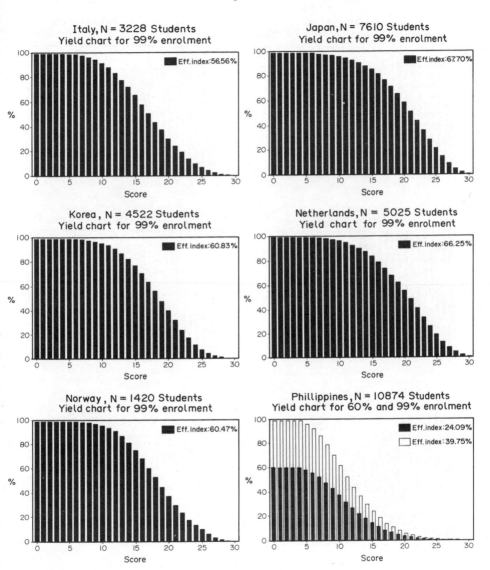

Figure 4 contd: Yield Histograms and Achievement Efficiency Indices for 17 Countries (Population 2)

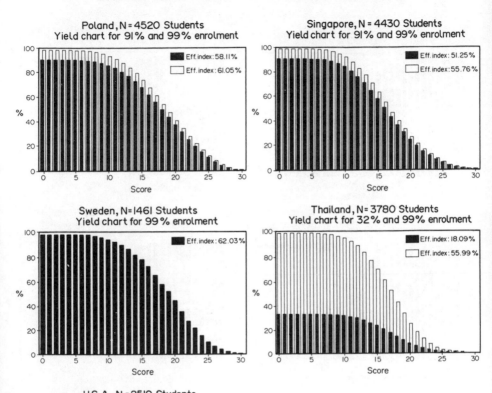

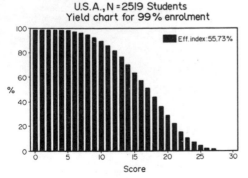

Figure 4 contd: Yield Histograms and Achievement Efficiency Indices for 17 Countries (Population 2)

school. The rank order of countries is exactly the same between the shaded histogram achievement efficiency indices and the mean scores. However, the rank order of countries does change slightly between the shaded histogram achievement efficiency index and the black histogram achievement efficiency index.

The shaded area histograms achievement efficiency indices for the total samples range from 40 percent in the Philippines to 72 percent in Hungary. As with the mean scores Hungary, Japan and the Netherlands are bunched at the top and England, Italy, Thailand, Singapore, the U.S.A. and Hong Kong near the bottom with the Philippines much below them.

The black area histograms achievement efficiency indices range from 18 percent in Thailand to 72 percent in Hungary (98 percent in school). It is in the more detailed report to be published in 1989 that an attempt will be made to examine why there are these differences in yield.

School Achievement

Table 7 presents the data on school means for Population 2. The standard deviations range from 1.2 in Finland to 3.7 in Singapore. The roh values range from 4 percent in Japan to 56 percent in Singapore. The Scandinavian countries have low values in keeping with their school policies. Hungary was the highest scoring country at Population 2 level. The lowest school mean score in Hungary was 14.8. The final column of Table 7 presents the percentage of schools in each country scoring less than the lowest school in Hungary . Policy makers often identify a few low scoring schools in their own countries and attempt to improve the performance of such schools by injections of personnel and material resources. However, an international comparison allows certain systems to observe that many schools in their systems would be considered "disadvantaged" - in this case, in Hungary. The relatively high percentage of low scoring schools in Italy,

Table 7:   School Achievement Score Data for Population 2
(30 item Core Test)

Country	N	Mean	SD	Min	Median	Max	Roh	Percent of school means below 14.8
Australia	233	17.8	2.2	11.2	18.5	23.2	.17	8
Canada (Eng)	209	18.6	2.0	10.4	18.6	23.7	.14	6
England	147	16.7	2.4	11.6	16.4	23.2	.19	19
Finland	90	18.5	1.2	13.6	18.6	21.5	.05	2
Hong Kong	110	16.4	2.5	10.2	16.7	22.3	.29	26
Hungary	99	21.7	2.5	14.8	22.0	27.7	.26	0
Italy	291	16.7	3.4	6.3	16.2	27.5	.39	37
Japan	199	20.2	1.3	14.7	20.1	24.0	.04	1
Korea	189	18.1	2.0	12.9	18.1	24.0	.15	5
Netherlands	224	19.8	3.7	8.9	19.2	26.7	.50	16
Norway	77	17.9	1.3	13.7	17.9	24.4	.02	1
Philippines	269	11.5	3.2	6.1	10.7	25.5	.48	87
Poland	201	18.1	3.2	11.4	17.8	27.8	.34	14
Singapore	185	16.5	3.7	9.5	16.4	26.0	.56	32
Sweden	69	18.4	1.7	13.8	18.5	23.1	.08	1
Thailand	96	16.5	2.1	11.0	16.5	22.1	.24	26
U.S.A.	119	16.5	2.9	8.2	16.4	25.2	.29	30

Singapore and the United States is of note. Again, the attempt to identify alterable factors which are associated with the differences in scores between countries, between schools within countries, and between students within countries will be one of the major aims of the forthcoming full report.

Conclusion

Population 2 in this report is of great importance because, for most countries, this is a point near the end of full-time compulsory education and the science achievement levels are one indicator of how scientifically literate the general public and work force will be. England, Italy, Singapore, the U.S.A., and Hong Kong are performing poorly and would appear to have grounds for concern. Hungary and Japan perform well. Japan has very small differences between schools and would appear, therefore, to have very good equality of educational opportunity for its students wherever they go to school in Japan. The same is true for Norway and Finland and Sweden. Australia has slipped in its relative standing since 1970 among the ten countries in the 1970 study.

However, the scientific literacy of the general population is one thing. The science achievement of the elite in a technological era is another. The achievement results of those specializing in science are reported in the next chapter.

CHAPTER 4: POPULATION 3 TEST SCORES

It will be recalled that Population 3 consists of science students in the terminal grade in school. This is Grade 12 except for Ontario in Canada (English), England, Hong Kong , Singapore and the technology track in Sweden where it is Grade 13. Population 3 is split into two groups: a) the group of those studying biology, chemistry or physics and these three sub groups are known as Populations 3B, 3C and 3P respectively; and b) Population 3N consisting of those students not studying science in the terminal grade.

The data presented in this chapter are: a) the achievement scores of each of these four sub groups on the core test, the only test that was taken by all groups; and b) the test scores of those studying biology on the biology test, those studying chemistry on the chemistry test, those studying physics on the physics test, and those not studying science on a special non-science test.

Before proceeding to the presentation of results there are several points to be noted. It will be recalled from Chapter 1 that there is no test relevance index for the Population 3N since they do not study science. All countries tested Population 3 except for the Netherlands and the Philippines. However, the Population 3 data for Korea and Thailand are not included because the checking of these data sets had not been completed by the time that this preliminary report had to go to press. These countries' results will be published in a future report. The proportion of an age group in each population is very different from country to country (see Table 1b) and this is why the percentage of the equivalent age group represented by each of the Population 3 groups is given in each table. Hong Kong tested each of

45

the last two grades separately. In each table, Form 6 (Grade 12) is presented first and Form 7 (Grade 13) second.

In Finland, of the students in the academic school (45 percent of an age group) all study biology. In Hong Kong, half of the biology students in the biology sample do not study biology. The Swedish Population 3 is composed of all students in the science track and all in the technological line (the proportion being 50:50) but those in the technological line do not study biology, even though they are included in the biology sample.

It will be noted that there are no core test results for the United States. The United States tested in 1983/1984 and again in 1986, when the core test at the Population 3 level was omitted. The special tests for science students were administered in 1986, but five items were omitted from the biology test, five from the chemistry test, and four items were omitted from the physics test. Hence, biology and chemistry tests have only 25 items that are the same as the other countries and the physics test has 26 items. The items which had been excluded were replaced by items from the core test. Hence, the scores are presented in percentage frequencies but it must be noted that the United States with 25 items is being compared with other countries with 30 items. The reduced number of items in percentage form will result in a reduced range.

In Australia, one item was omitted from the biology test scoring because of a typing error in that item in the test which rendered the item invalid. In Canada (English) one item had to be removed from the physics test for the same reason. Because the scores had to be calculated in percentage terms, it must be remembered that, for most countries, the percentage is out of thirty items, but for the biology test in Australia and the physics test in Canada it is out of twenty-nine items, and for the United States it is out of twenty-six items (Physics) and twenty-five items (Biology and Chemistry).

The weighting of the samples at Population 3 and 3N levels was a problem. In some countries, it was possible to obtain statistics or estimated statistics on the number of students in the target population for each stratum of those studying biology, chemistry, and physics but not for Population 3 overall, that is, for those studying science. For Population 3N it was usually difficult to obtain estimates for the number of students in the target population for each stratum that were not currently studying science. Stratum weights could be calculated for each group separately in countries where target population and design sample figures existed, but in other countries, stratum weights were calculated using the total Grade 12 (or 13) enrolment figures for all students.

Core Test Results

Table 8 presents the scores of the science students and the non science students on the core test. The scores were weighted using either school or stratum weights. In general, the chemistry and physics students score higher than the biology students who are followed by the non science students. The highest score of all is achieved by the English chemistry students. In general, Hong Kong, England, Hungary, and Japan (except for their biology students) achieve high scores. Hungary's non science students also score very well compared with the non science students in other countries. Italy's science students have the lowest scores. The differences in score points between the lowest and highest achieving countries are 8.5, 7.7, 7.9, and 8.8 for the biology, chemistry, physics, and non science groups respectively. As will be seen in Chapter 5 one year of schooling represents about 2 score points, so that the differences in scores on the core test represent about 4 years of schooling.

In interpreting the achievement score differences, it is, of course, necessary to bear in mind the age group represented by each of the country populations in the table. Information on the differences in age and percentage in school are given in Table 1b and in Tables 10 to 13.

All of the students taking the biology, chemistry, and physics tests

Table 8:   Core Test Scores of Science Students and Non-Science Students (Population 3)

Country	Biology Students			Chemistry Students			Physics Students			Non-Science Students		
	N	Mean	SD	N	Mean	SD	N	Mean	SD	N	Mean	SD
Australia	1631	18.0	4.7	1177	22.7	4.4	1073	22.2	4.7	995	15.1	4.8
Canada (Eng)	3254	18.0	5.4	2923	20.1	5.2	2766	20.6	5.1	509	13.2	4.9
England	884	23.3	4.3	892	25.5	3.4	917	24.6	3.6	1004	18.2	4.8
Finland	1652	19.5	4.8	971	20.3	4.9	810	22.0	4.1	-	-	-
Hong Kong (Form 6)	5960	23.6	3.4	6018	23.6	3.4	6025	23.6	3.4	-	-	-
Hong Kong (Form 7)	3614	25.0	3.1	3670	25.0	3.1	3679	25.0	3.1	-	-	-
Hungary	301	23.7	3.8	143	24.4	3.8	398	25.2	3.6	1036	22.0	4.7
Italy	147	16.5	5.2	217	17.8	5.7	1766	17.3	5.2	2455	13.9	5.0
Japan	1212	19.5	5.7	1468	24.0	5.0	1187	24.3	4.3	2229	18.0	5.8
Norway	276	19.6	4.9	283	20.8	4.9	443	22.4	4.6	595	14.7	4.4
Poland	764	20.6	4.6	765	20.5	4.6	1716	21.5	4.4	-	-	-
Singapore	902	24.9	3.3	945	22.8	3.9	1071	21.9	3.9	297	15.7	4.8
Sweden	1232	22.7	4.2	1172	22.9	4.1	1156	23.0	4.0	-	-	-

were combined and their results on the core test calculated. Since there are no non science students included, the achievement may be regarded as that of all students "specializing in science" on a general science test. Given the complex structure of the sampling at Population 3 level it was not, at this stage, possible to calculate appropriate weights for performance on the core test and, therefore, the data in Table 9 are unweighted.

Since, the core test taken by students at Population 3 level contained 15 items which were common to items in the core test at Population 2 level, it is the scores on these common items which were used in the calculation of growth scores in Chapter 5. This required the calculation of scores not only on the common items but also on the total core test. Table 9 presents these results.

Even though the growth scores so calculated are less than perfect measures of the changes in performance across the terminal years of schooling, it is believed that a reasonable estimate has been made of the magnitude of increase in level of achievement for all science students as scanning of the data in Tables 8 and 9 will reveal.

Table 9: Student Achievement Score Data for Population 3 Science Students on Core Test (unweighted)

Country	N	Mean	SD	25%	Median	75%	Age
Australia	3881	20.7	5.1	17	21	25	17:3
Canada (Eng)	8943	19.3	5.4	16	20	23	18:3
England	2693	24.6	3.8	22	25	27	18:0
Finland	2552	20.0	4.9	16	20	24	18:7
Hong Kong (Form 6)	6025	23.7	3.4	22	24	26	18.4
Hong Kong (Form 7)	3679	25.1	3.0	23	25	27	19.2
Hungary	842	24.4	3.8	22	25	27	18:0
Italy	2130	17.4	5.2	14	18	21	19:0
Japan	3867	22.7	5.5	19	24	27	18:2
Norway	1002	21.1	4.8	18	22	25	18:11
Poland	3245	21.0	4.5	18	22	24	18:7
Singapore	2918	23.1	3.9	21	24	26	18:1
Sweden	2430	22.8	4.2	20	24	26	19:0

50   Science Achievement in Seventeen Countries

The top scoring country is Hong Kong Form 7 with approximately 8 percent of an age group being science students. England and Hungary with 5 and 3 percent of an age group respectively in this population follow closely. Italy, Canada (English) and Finland have relatively low scores. The mean level of achievement on this general test was high and the range was from 17 to 25 score points out of a possible 30.

England, Hong Kong, and Singapore, in contrast to their performance at Population 1 and 2 which was low, have a very high achievement level at Population 3. Is it their policy to give special priority to their elite at the top of the system and less attention to the education of the general population when nearly one hundred percent of an age group is in school? Finland has reformed its comprehensive school system which includes Populations 1 and 2 but has not taken specific measures for the improvement of the terminal grade of secondary school since the examination at the end of the terminal grade had been expected to "maintain standards" at that level. However, at the end of 1986 a special committee was established for the reform of science education. The results presented above will be of interest to this committee.

Science Students' Test Scores

The performance of the science students on the specific biology, chemistry, and physics tests are likely to be of more interest than their performance on a general test. All scores are weighted in the special science tests.

Tables 10 to 12 present the scores of the students studying biology, chemistry and physics on their respective subject tests.

Table 10 presents the scores for the biology students on the biology test. The student means and standard deviations are presented and these are followed by the roh values, the percentage of schools scoring less than

Table 10: Biology Students Achievement (Population 3)

Country	N students	Mean	SD	Roh	% of schools scoring below 56.6	% in course	Age
Australia	1631	48.2	13.9	.10	93	18	17:1
Canada (Eng)	3254	45.9	14.0	.20	95	28	18:2
England	884	63.4	13.1	.20	14	4	18:0
Finland	1652	51.9	12.8	.04	91	45	18:7
Hong Kong (Form 6)	5960	50.8	14.8	-	77	7	18:4
Hong Kong (Form 7)	3614	55.8	16.8	-	47	4	19:2
Hungary	301	59.7	13.5	.38	37	3	18:0
Italy	147	42.3	14.1	.28	100	14	19:5
Japan	1212	46.2	15.1	.39	84	12	18:1
Norway	276	54.8	15.0	.07	56	10	18:11
Poland	764	56.9	12.9	.32	48	9	18:8
Singapore	902	66.8	12.8	.11	0	3	18:0
Sweden	1232	48.5	15.8	.18	78	15	19:0
U.S.A.	659	37.9	15.4	.40	98	6	17:5

the lowest school in Singapore(56.6%), the percentage of an age group in school and the average age of those studying biology. The values range from a low of 38 percent in the U.S.A. to 67 percent in Singapore. The rank order correlation between the national average age and national scores is close to zero. The roh values are high in the United States, Japan, and Hungary and the fact that over 90 percent of schools in Italy, Finland, the United States, Canada (English), and Australia score below the lowest school in Singapore is worth noting. It should be remembered, however, that all students in the academic stream in Finland take biology, and that half of the students in the samples in Hong Kong and Sweden do not study biology.

Table 11 presents the scores for the chemistry students on the chemistry test and the type of data presented are the same as for Table 10. Finland has the lowest score and Hong Kong Form 7 the highest. Only Finland and the United States have large percentages of schools scoring below the lowest school in Hong Kong (30.0%).

Table 11: Chemistry Students Achievement (Population 3)

Country	N students	Mean	SD	Roh	% of schools scoring below 30.0	% in course	Age
Australia	1177	46.6	18.8	.20	4	12	17:3
Canada (Eng)	2923	36.9	16.0	.25	26	25	18:4
England	892	69.5	17.2	.21	0	5	18:0
Finland	971	33.3	13.7	.12	24	14	18:6
Hong Kong (Form 6)	6018	64.4	17.0	-	0	14	18:4
Hong Kong (Form 7)	3670	77.0	17.4	-	0	8	19:2
Hungary	143	47.7	18.3	.43	13	1	18:1
Italy	217	38.0	23.4	.60	33	2	19:3
Japan	1468	51.9	22.0	.62	14	16	18:2
Norway	283	41.9	16.8	.12	4	15	18:11
Poland	765	44.6	17.1	.43	13	9	18:7
Singapore	945	66.1	17.4	.28	0	5	18:0
Sweden	1172	40.0	16.6	.17	11	15	18:11
U.S.A.	537	37.7	18.3	.49	48	1	17:8

Table 12 presents the scores for the physics students on the physics test. The format of presentation is the same as for biology and chemistry. Again, the two Hong Kong forms have the highest scores. It is interesting that Form 6 which is equivalent to England's first year sixth form should perform at about the same level as England's second year sixth form. Italy is particularly weak. The roh value for Hungary is particularly high indicating large differences between schools. The percentages of schools in some countries scoring lower than the lowest school in Hong Kong are large. The United States which has a much smaller percentage of an age group than Hong Kong studying physics has 89 percent of its schools scoring lower than the lowest school in Hong Kong.

What can be said about the student performance at Population 3 level? England, Hong Kong and, to some extent, Singapore have high achievement levels for Population 3 despite the fact that they performed poorly at Populations 1 and 2 levels. This may, in part, be explained by the fact that "hard" teaching in science begins only in the third form of secondary

Table 12: Physics Students Achievement (Population 3)

Country	N students	Mean	SD	Roh	% of schools scoring below 50.5	% in course	Age
Australia	1073	48.5	15.1	.15	67	11	17:3
Canada (Eng)	2766	39.6	14.6	.23	93	19	18:4
England	917	58.3	14.9	.12	18	6	18:0
Finland	810	37.9	13.8	.10	93	14	18:6
Hong Kong (Form 6)	6025	59.3	14.7	-	11	14	18:4
Hong Kong (Form 7)	3679	69.9	14.4	-	0	8	19:2
Hungary	398	56.5	17.2	.42	45	4	18:1
Italy	1766	28.0	12.9	.37	99	19	19:3
Japan	1187	56.1	17.2	.42	36	11	18:2
Norway	443	52.8	15.6	.12	44	24	18:11
Poland	1716	51.5	17.2	.46	53	9	18:7
Singapore	1071	54.9	13.2	.07	25	7	18:0
Sweden	1156	44.8	14.9	.08	82	15	18:11
U.S.A.	485	45.5	15.8	.38	89	1	17:8

school. In addition, a self selection process takes place at about age 16 and there is a concentration of learning in the two last years of secondary school when the majority of those studying science will typically be studying a combination of 3 subjects only in science and mathematics.

Canada (English), Finland, Italy and the U.S.A. perform relatively poorly at Population 3 level. Italy and the U.S.A. did not perform well at lower levels of schooling and these two countries must be concerned about their science education in general. Canada (English) and Finland both have unexpectedly lower performance at Population 3 level compared with their performance at Population 2 level. Japan and Hungary perform relatively well in their specialist populations except for Japan in biology. If selectivity is measured by the percentage of an age group studying the specialist subjects in the terminal year of secondary education, having a lower percentage studying the subject does not necessarily result in higher achievement. Similarly, the average age of those studying science does not relate positively with achievement.

The roh values for Japan are very low for Population 1 and 2 but very high for Population 3. At the lower population levels there are very few private schools at Grade 12 level. However, about one quarter of the schools are private. Some of the private schools are famous for their elite education. There are also other private schools which take students who have failed the examation to enter senior high school in the public system. Hence, the differences among schools in the ability of students is very high.

Non Science Students' Test Scores

Eight countries tested their students not studying science on a special 30 item science test. Table 13 presents the results. The means and standard deviations are for students. Roh is also presented and the percentages of schools scoring at a lower level than the lowest school in Hungary (14.8) are given. The English non science students perform well but Canada (English) and Italy have relatively low scores. The school differentiation is high in Italy. Although the ages are not presented for this population the students tend to be two or three months older than their counterparts studying science.

Table 13:   Non-Science Students Achievement (Population 3)

	N Students	N Schools	Mean	SD	Roh	% of schools scoring below 14.8
Australia	995	120	16.4	4.4	.17	21
Canada (Eng)	509	81	14.9	4.5	.22	42
England	1004	126	18.1	4.0	.11	5
Hungary	1036	77	18.9	3.9	.26	0
Italy	2455	119	15.0	4.4	.36	40
Japan	2229	60	17.3	4.0	.30	22
Norway	595	78	16.1	3.9	.13	26
Singapore	297	9	15.6	3.8	.14	33

Conclusion

In the science students' results there are considerable differences: Singapore and England achieve high scores in Biology but Hong Kong and England have the highest scores in Chemistry and Physics. Canada (English), Finland, Italy, and the United States tend to perform poorly. Hungary and Japan tend to perform relatively well. Japan, however, achieves at only a relatively low level in Biology. Several countries have high proportions of schools scoring lower than the lowest scoring schools in the highest scoring countries. In particular, Australia, Canada (English), Finland, Italy, Sweden, and the United States have many poor scoring schools and it would seem likely that the school authorities would have grounds for concern.

CHAPTER 5: SPECIAL ANALYSES

This chapter presents some small analyses based on the data presented earlier in this report. The first concerns the growth from Populations 1 to 2 and from Populations 2 to 3 in the different countries. Comber and Keeves (1973)[1] presented growth data and their figures from the first science study became one of the most quoted figures in international educational research. The second analysis concerns sex differences in science achievement to what extent they exist and whether the differences increase from population level to population level. The third concerns relationships between curriculum indices and achievement. The fourth involves relationships between achievement at the terminal grades of secondary school and the percentage of the grade cohort in school. The word 'growth' is used in this context to denote the increase in achievement from one level in a school system to another.

Growth

Table 14 presents the means and standard deviations on the anchor items for Populations 1, 2 and 3. There were 13 items which were common to Populations 1 and 2 and 15 items which were common to Populations 2 and 3. It can be seen that from Population 1 to Population 2 the smallest gain (1.5) was made in the Philippines which also had the lowest score at Population 1 level. The largest gain (3.4) was made in Hong Kong which also had a relatively

---

1   Comber, L.C., and Keeves, J.P., (1973) Science Education in
     Nineteen Countries. New York. John Wiley.

5 7

Table 14: Anchor item scores between populations

	13 items				15 items								
	Pop.1		Pop.2		Pop.2		Pop.3 Science Students		Pop.3 Non-Science				
	Mean	SD	Mean	SD	Mean	SD	Mean	SD	Mean	SD			
Australia	6.6	2.8	9.9	2.4	6.8	2.9	11.0	2.7	8.5	2.8			
Canada (Eng)	7.3	2.7	10.1	2.2	7.3	2.9	10.8	2.9	7.9	3.0			
England	6.1	2.9	9.3	2.5	6.3	2.9	12.9	1.9	9.9	2.7			
Finland	8.1	2.5	10.2	2.0	7.0	2.6	10.7	2.7	-	-			
Hong Kong (Form 6)	5.9	2.6	9.3	2.3	6.1	2.6	12.6	1.8	-	-			
Hong Kong (Form 7)	-	-	-	-	6.1	2.6	13.0	1.7	-	-			
Hungary	7.8	2.7	10.9	2.0	9.8	3.1	12.9	2.0	11.6	2.6			
Italy	6.8	2.8	8.8	2.5	6.9	2.9	10.5	2.9	8.6	3.1			
Japan	8.0	2.4	10.1	2.1	9.0	3.3	12.2	2.8	9.9	3.2			
Korea	7.8	2.6	9.4	2.2	7.3	2.8	-	-	-	-			
Norway	6.6	2.6	9.6	2.2	7.0	2.9	11.2	2.6	7.9	2.6			
Philippines	4.7	2.6	6.2	2.7	4.4	2.3	-	-	-	-			
Poland	6.2	2.6	9.3	2.3	7.8	3.2	12.0	2.5	-	-			
Singapore	5.8	2.5	9.1	2.3	6.2	2.9	12.2	2.1	8.8	2.8			
Sweden	7.9	2.6	10.1	2.3	7.3	2.9	12.1	2.2	-	-			
U.S.A.	6.9	2.8	9.1	2.7	6.2	2.7	-	-	-	-			

- not available

low score at Population 1 level. Hungary's gain (3.1) was larger than Japan's (2.1) but Japan's score was the highest at Population 1 level.

From Population 2 to Population 3 science students, the highest gains were for Hong Kong Form 7 (6.9), England and Hong Kong Form 6. This reflects the high degree of specialization mentioned in Chapter 4. In the Population 3 non-science students, the highest gain was in England and the lowest in Canada (Eng).

Since the anchor items were contained in the Population 3 core test and the United States did not administer this test, it was not possible to calculate growth scores.

It is possible to employ these growth scores to develop a common standard score scale. Performance on the core tests at each population level can in turn be converted into standard scores and performance on the core tests rescaled using the common standard score scale. Thus, all countries' mean scores can be brought onto one scale (see Appendix 10 and Comber and Keeves, 1973, pp 167-168 for further explanation). This procedure does not provide a measure of absolute gain on common items from one population to the next as given by the growth scores on the anchor items in Table 14. Rather it presents the relative performance of each country at the different levels as recorded on a single common scale in such a way that differences between countries and population levels can be examined. The standard scores and growth scores for all countries are given in Appendix 10.

Figure 5 presents in diagrammatic form the relative performance of each of the three population levels on the common scale as well as the growth in performance between age or grade levels for each country. The results presented in Appendix 10 and Figure 5 show the effects of providing a firm foundation for the study of science through the systematic teaching of science in the primary school, where Japan and Korea would appear to excel with standard scores of (-0.67) and (-0.69) respectively.

Growth scores from Population I to Population 2 show the consequences of specialization in science during the lower secondary school

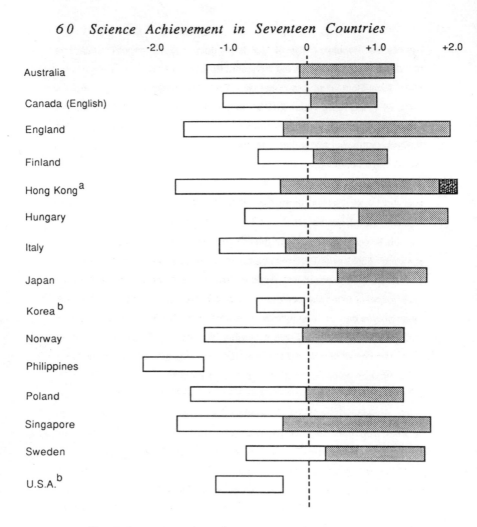

Figure 5:   Increase in Level of Achievement in Science from
Population 1 to Population 2 and Population 3
(Science Students)

a   increase and performance in Hong Kong from Form 6 to Form 7
shown by ▒▒▒

b   further analysis required to obtain results for Korea and the U.S.A.

grades, commonly by the intensive teaching of science at this level. Here Hungary, with a growth score of 1.63 excels, and is followed by Poland (1.57) working from a lower level of achievement at the primary school level and by those countries following the English science tradition, Singapore (1.41), Hong Kong (1.40) and England (1.34). Korea (0.64) and Finland (0.75) after relatively high initial performance at the primary school level do not appear to maintain an emphasis on science in the lower secondary school curriculum although still performing creditably at this level. The three countries with an English science tradition of specialization in science at the upper secondary school level, maintain growth from Population 2 to the Population 3 (science students) level, with England (2.24), Hong Kong (2.16) and Singapore (2.03) recording the highest levels of growth among the countries listed.The relationships between growth from Population 2 to Population 3 (science students) and participation rates in science and the particular science subjects and retention rates at school warrant thorough examination.

Much interesting information is available from a careful consideration of these growth scores. For example, although Norway at no population level records a high measure of performance, it would appear to show a relatively strong level of growth across the years of compulsory schooling, in part as a consequence of the low level of achievement at Population 1, and in part as a consequence of testing a higher grade level at Population 2. Sweden, with a high level of achievement at Population 1, and relatively strong achievement at later population levels, shows less growth than Norway from Population 1 to Populations 2 and 3, but more growth from Population 2 to Population 3 because more years of schooling are involved. Likewise, Finland with a high level of achievement at Population 1, records significantly lower levels of growth across the years of schooling. It should be noted from the evidence in Appendix 9 that only in England and Hungary do the Population 3 non-science students record strong growth from Population 2 to Population 3. It would be of interest to examine whether this is a consequence of the taking of non-specialist science courses covering all three branches of science:

biology, chemistry and physics, or merely a consequence of the selection of students of superior ability to continue with their studies at the pre-university level.

One standard score in Figure 5 is the equivalent of just under four years of schooling (based on an international average) between Populations 1 and 2, and about three years of schooling between Populations 2 and 3. It is, therefore, possible to make comparisons of which grade level in one country is equivalent to which grade level in another country. Thus, Grade 8 in the United States has about the same level of achievement as Grade 6 in Japan and Korea. It is left to the reader to make further comparison of interest to him or her.

Sex Differences in Science Achievement

In the first international science study in 1970 and 1971 boys consistently performed better than girls in all countries and the gap increased as students ascended the school system. The difference was presented as a standard score for each country computed by subtracting the mean score for girls from the mean score for boys, and dividing this figure by the mean standard deviation for all countries. The standard score difference for achievement in science went from 0.23 at Population 1 to 0.46 at Population 2 to 0.69 at Population 3.

In the Second Science Study, not all countries tested the non-scientists at Population 3 level. Therefore, only the sex differences for the specialist tests are given for Population 3. This is not strictly comparable to the results from the first science study.

In examining these standard score differences it should be noted that only values greater than the following should be regarded as being significant and worthy of note: Population 1: 0.17; Population 2: 0.18; Population 3 Biology: 0.20; Population 3 Chemistry: 0.34; Population 3 Physics: 0.21, and Population 3 Non-Science: 0.20.

Table 15 below presents, for each country in Population 1, the mean score for boys, the mean score for girls, the difference of these two mean scores as a result of subtracting the girls' score from the boys' score and the standard score difference by dividing the difference in mean scores by the mean standard deviation (4.30).

Table 15: Sex Differences in Science Achievement (Population 1)

	Boys	Girls	Difference	Standard Score Difference
Australia	13.44	12.26	1.18	.27
Canada (Eng)	14.20	13.24	.96	.22
England	12.12	11.27	.85	.20
Finland	15.96	14.62	1.34	.31
Hong Kong	11.60	10.78	.82	.19
Hungary	14.78	14.11	.67	.15
Italy	13.78	13.05	.73	.17
Japan	15.56	15.29	.27	.06
Korea	16.05	14.61	1.44	.33
Norway	13.67	11.78	1.89	.44
Philippines	9.60	9.41	.19	.04
Poland	12.51	11.47	1.04	.24
Singapore	11.80	10.66	1.14	.27
Sweden	15.15	14.13	1.02	.24
U.S.A.	13.79	12.54	1.25	.29

The mean score point difference is 0.91 and the mean of the standard score differences is 0.23, showing a superiority of boys over girls. Norway, Korea, and Finland have the largest differences and the Philippines and Japan the lowest.

Table 16 presents similar data for Population 2. The mean standard deviation is 4.74.

From Table 16, it can be seen that the average standard score difference in achievement between the sexes is 0.34 and this is lower than the 1970 figure of 0.46. Hungary has the lowest difference between boys and

Table 16:   Sex Differences in Science Achievement
(Population 2)

	Boys	Girls	Difference	Standard Score Difference
Australia	18.44	17.12	1.32	.28
Canada (Eng)	19.51	17.69	1.82	.38
England	17.59	15.87	1.72	.36
Finland	19.18	17.81	1.37	.29
Hong Kong	17.17	15.41	1.76	.37
Hungary	22.18	21.49	.69	.14
Italy	17.43	15.95	1.48	.31
Japan	20.92	19.43	1.49	.31
Korea	18.98	17.01	1.97	.41
Netherlands	20.96	18.65	2.31	.49
Norway	18.85	16.99	1.86	.39
Philippines	11.92	11.11	.81	.17
Poland	18.91	17.34	1.57	.33
Singapore	17.38	15.51	1.87	.39
Sweden	19.28	17.50	1.78	.38
Thailand	17.39	15.76	1.63	.34
U.S.A.	17.36	15.53	1.83	.39

girls and the Netherlands the highest. However, the 1970-71 test contained
one quarter of the total number of items that assessed performance on
practical work through a pencil and paper practical test. This type of item, which
had a strong bias in favour of male students, was not included in the 1983-84
tests and could well account for the reduction in the standard score sex
differences reported above. More detailed analyses are clearly warranted.

Table 17 presents the standard score differences for the science
students and non-science students between boys and girls on the biology,
chemistry, physics and non-specialists tests as appropriate. The biology
students had a standard score difference of 0.17, the chemistry students of
0.36, the physics students of 0.39 and the non-science students of 0.34. The
values recorded for each country are influenced greatly by the relative
proportions of the age cohort which study each subject at the terminal
secondary school stage. No comparable data were reported from the First IEA

Table 17: Standard Score Differences in Science Achievement for Boys and Girls (Population 3)

| | **Scores Student Tests** | | | |
	Non-Science	Biology	Chemistry	Physics
Australia	.27	-.02	.48	.24
Canada (Eng)	.25	.28	.40	.43
England	.28	.19	.32	.04
Finland	-	.19	.42	.50
Hong Kong (Form 6)	-	.01	.34	.45
Hong Kong (Form 7)	-	-.18	.24	.45
Hungary	.32	.18	.70	.40
Italy	.51	.39	.12	.32
Japan	.42	.20	.34	.30
Norway	.47	.44	.41	.38
Poland	-	.17	.44	.59
Singapore	.18	.49	.41	.47
Sweden	-	-.23	.04	.39
U.S.A.	-	.33	.42	.44
Average	.34	.17	.36	.39

Science Study for such groups of students. It should be noted that the girls outperform the boys in biology in two countries: Hong Kong Form 7 (0.18) and Sweden (0.23), at this stage of schooling.

Curriculum Indices and Achievement

It is now of interest to relate the levels of achievement in science on the different tests to the three curriculum indices presented and discussed in Chapter 1 and Appendix 8. In Table 18, the product moment correlations of the three curriculum indices, the curriculum relevance index, the test relevance index, and the curriculum coverage index, with achievement are presented for Populations 1 and 2 and only for the chemistry and physics groups of students for Population 3. There was insufficient variation between countries in the values of the curriculum indices for Population 3 biology students to allow meaningful correlations to be calculated.

The negative correlations at Population 1 level for all three curriculum indices are perhaps not surprising. They result primarily from the high level of achievement in Japan which has low curriculum coverage, and the low level of achievement in the Philippines, with relatively high curriculum coverage. However, they also suggest that some of what has been learnt about science by the age of 10 years has been acquired from informal sources, including magazines, the media and visits to museums.

The positive correlations between the three curriculum indices and achievement at Population 2 level are to be expected as a consequence of formal instruction in science during the years in the lower secondary schools in all countries. It is clear, however, that the differences in emphasis on science result in different levels of achievement across countries. While the correlations are not high, they are consistent with expectation in spite of the relative crudeness of the measures employed for the curriculum indices.

No relationship is found between the curriculum coverage index and achievement on the physics test at Population 3, and a negative relationship for chemistry at this level. The low variability between the values of the curriculum coverage indices across countries inevitably leads to such results. However, it should be noted that the values of the correlations for the curriculum relevance index and the test relevance index are positive as might be expected.

It is clear from these results for Populations 2 and 3 that where the items in the tests are more relevant to the curricula of the different countries the level of achievement is generally higher and where the items are less relevant to the curricula, achievement is lower. However, as indicated in Chapter 1 and Appendix 9, there is not, in the main, great variability between the values of the indices at each population level; hence, the relatively low positive correlations. At Population 1 level the apparent effects of informal learning in science are of some consequence.

In conclusion, it is important to emphasize that all the indices calculated here are based on the intended curriculum in centralized school systems and on estimates obtained from samples on what is taught in those countries which do not have a national curriculum or syllabus. Further analyses need to be carried out between measures of opportunity to learn and achievement to examine the discrepancies which clearly exist between what is intended and what is achieved. It is only through consideration of opportunity to learn and through multivariate analyses which examine, in particular, what teachers teach and do that the mismatch between intention and achievement can be explained. It is the major task of the further publications to present the results of such analyses.

Percentage of an Age Group in School and Achievement

The correlations across the thirteen countries between the percentage of an age group studying the different sciences at the terminal secondary school level and achievement in the three branches of science are -0.34 for biology,-0.41 for chemistry and -0.16 for physics. It would appear that the lower the proportion of an age group in school in a country and studying a branch of science, the higher the level of achievement of the students in that country. However, it should be noted that the correlations recorded are at best only moderate and are not high, suggesting that factors other than the ability of the students remaining at school and engaged in the study of science are involved. Such factors are likely to include the extent of specialization both at the lower secondary school level and, in particular, in the terminal year of schooling, which would be reflected in the time given to the study of science and the number of subjects studied at this level.

It should be noted by reference to Table 1 in Chapter 1 and to Table 11 in Chapter 4 that in chemistry, for which the largest correlation is reported above, countries such as Sweden (15 percent of an age group), Norway (15 percent of an age group) and Hong Kong (14 percent of an age group) score considerably higher than the United States with only one percent of the age group studying advanced placement chemistry. Similar comments could be made with respect to other countries for their levels of achievement in biology and physics. If some countries score higher with a substantial proportion of an age group studying a science subject, than other countries with a lower percentage of an age group, it is important that further analyses should be undertaken to examine these disparate cases in detail.

Conclusion

Achievement at each population level was brought on to a common scale in order to examine the growth in achievement from one level of schooling to another. England, Hong Kong and Singapore have relatively very high growth especially from Population 2 to Population 3 levels, where the groups are highly selected and study only three subjects (typically a combination of science and mathematics). One standard score in Figure 5 is the equivalent of just under four years of schooling between Populations 1 and 2 and about 3 years of schooling between Populations 2 and 3. The Philippines Grade 9 achievement is the same level as Grade 4 in Norway and much lower than Grade 5 achievement in Japan and Korea. The difference between Grade 9 score in the Philippines and Grade 5 score in Japan is the equivalent of about two years of schooling. Grade 9 in the United States has about the same level of achievement as Grade 7 in Japan and Korea. Grade 12 in Italy has about the same achievement level as Grade 8 in Hungary.

The differences in achievement between boys and girls is generally in favour of boys at all population levels. At Population 1, the difference is 0.23 of a standard deviation, at Population 2 it is 0.34, and at Population 3 it is 0.17 for biology (with girls outperforming boys in Sweden and Hong Kong), 0.36 for chemistry, and 0.39 for physics.

An examination was made of the relationships between the three curriculum indices and achievement at each population level. The curriculum indices are based on rough estimates of emphasis in the curriculum in each country and on the extent to which particular topics are meant to be taught in school, that is, of the intended curriculum. The correlations at Population 1 are negative and their negative values are due primarily to Japan having high achievement and curriculum indices being low, and, on the other hand, the achievement for the Philippines being low and the indices high.

For Population 2 there are moderate positive correlations and for Population 3 the same is true except for the curriculum coverage index. The low correlations between the intended curriculum and achievement indicate that much depends on how the curiculum is implemented by the schools and teachers. This will be the focus of attention in further publications.

Finally, the relationships between achievement and the proportions of an age group studying each of the science subjects at Population 3 were examined. In general, the correlations were negative, indicating that the smaller the percentage of an age group, the higher was the achievement. But, care must be exercised in interpreting this finding since only thirteen countries were involved and because there are some cases where a country with a higher proportion of an age group in school achieves at a higher level than a country with a small proportion of an age group. For example, Sweden, Norway, and Hong Kong with 15 percent, 15 percent and 14 percent of an age group achieve at a higher level in chemistry than the United States with only 1 percent of an age group in advanced placement chemistry.

CHAPTER 6: CONCLUSIONS

The data presented in this report for Populations 1 and 2 are all based on the core tests given in each country at each population level: a 24 item test for Population 1 and a 30 item test for Population 2. In the main reports to be published in 1989 it is anticipated that achievement based on all tests will be reported but it is expected that the overall picture will be much as given here, since the correlations among the core and rotated tests are very high. In one of the planned reports the results of all participating countries will be given with data analyzed to explain both between country performance differences and between student within country differences. Extensive analyses will also be undertaken of the differences in achievement between 1971-71 and 1983-86 and the factors associated with the achievement levels on the two occasions for the ten countries reported in this booklet.

From this initial examination of the performance results in the seventeen countries in the mid 1980s the main points of interest are summarized below.

Mass Education

Population 1 represents students towards the end of the first stage of education, usually termed primary or elementary education, 100 percent of an age group is enrolled in school. Population 2 represents the initial stage of secondary education. With the exception of the Philippines and Thailand, the countries have over 90 percent of an age group enrolled in school.

Finland, Hungary, Japan and Sweden perform well at these two levels. England, Hong Kong, Singapore and the U.S.A. perform relatively poorly with the U.S.A. descending in rank order from the 10 to 14 year old level. British primary school science has had a high reputation and its efficacy must be questioned in the light of these results.

The relative increase in achievement from Population 1 to Population 2 in Hungary and Poland is noteworthy. One question to be pursued in later analyses, therefore, is what happens in science education (or education in general) in these two countries that does not occur in other countries. The difference in the performance between boys and girls is also least in Hungary at the Population 2 level.

Population 2 is at a point, in most educational systems, which is just before the end of full time compulsory education. The amount of scientific literacy can be assessed through 'yield' coefficients. Hungary and Japan have high yields but England, Hong Kong, Italy, Singapore, and the U.S.A. have low yields. For those in school, Thailand and the U.S.A. have similar yields but if the yield of the proportion of an age group in school is calculated, Thailand's yield is much smaller. The Philippines has a low mean score and low yield. In October, 1987, the Philippines government doubled the education budget and it is to be hoped that educational achievement will improve. What is clear is that countries with low percentages of an age group in school cannot expect to increase their yield until a much higher percentage of an age group is enrolled in secondary education.

England, Hong Kong, Italy, Singapore, and the U.S.A. must be concerned about the scientific literacy of their general work force unless they remedy the situation through training programs or vocational education at a later stage.

'Elite' Education

Population 3 is the terminal grade of secondary school. As can be seen from Table 1 the proportion of an age group enrolled in academic school at the end of secondary school range from 17 percent in Singapore to 79 percent in the U.S.A. It is of interest to note that England, Hong Kong, and Singapore have kept their enrolments low at 20, 20, and 17 percent respectively. Japan has only 63 percent enrolled in academic streams but has another 31 percent enrolled in vocational courses.

Population 3 represents for most countries the development of an elite in science for continuation to the university and leading positions in industry. Hong Kong, England, and Singapore together with Hungary and Japan would appear to be educating their elite relatively well. However, Canada (Eng), Italy, and the United Sates would appear to have grounds for concern unless the situation is remedied at the university level. The very high proportion of schools in some countries scoring lower than the lowest school in the highest scoring country is dramatic and it is imperative that many detailed analyses be undertaken to identify the reasons for this.

Further Comments

Comments on each country's achievement have been given in the executive summary at the beginning of this preliminary report. Occasionally, remarks have been made about countries changing in rank order since the 1970 first science study was conducted. Appendix 11 presents the rank ordering of countries on those two occasions.

Finally, Figure 5 in Chapter 5 allowed all populations' achievement to be compared on one scale. Some of the differences are large and there are cases of the 10 year olds in two countries (Japan and Korea) scoring at a higher level than 14 year olds in another country (Philippines), and of 14 year olds in one country (Hungary) scoring higher than Grade 12 level in another (Italy). These disparities, representing several years of schooling, must give cause for concern.

As already pointed out several times, work will now proceed on examining the relationship of many of the student, teacher, school and system variables with achievement in order to identify those alterable variables which significantly effect the variation in achievement between and within countries. Information exists in the data files on some five hundred variables (data such as the amount of time spent studying science, the amount of homework and practical work, the structure of science education - integrated, separate subjects or "layer-cake" approach - the training and retraining of science teachers, the number of subjects studied in addition to science, what the teacher actually teaches, the size of the school and class etc.). Given that research funding is forthcoming, the results of these analyses will be included in the publications planned for release in 1989.

APPENDICES

APPENDIX 1: Definitions of National Target Populations

Population 1

Australia
All students of age 10:0 to 10:11 years in normal[1] schools in Years 4, 5 and 6 on
31 October 1983.

Canada (English)
All students in Grade 5 on 1 May 1984.

England
All students in normal schools in Year 5 in the age range 10:0 to 10:11 on
1 September 1984; that is, of age 9:9 to 10:8 on 1 June 1984.

Finland
All students attending Grade 4 classes in regular Finnish-speaking comprehensive
schools on 15 April 1984.

Hong Kong
All students at the fourth year of primary education studying science as part of the
Hong Kong Primary School curriculum on 31 May 1984.

Hungary
All students in the 4th class on 20 May 1983.

Italy
All students enrolled in the 5th grade of State Elementary Schools on
20 April 1983.

Japan
All students in Grade 5 of primary schools on 15 May 1983.

Korea
All students in the 5th grade on 11 November 1983.

Norway
All students of age 10:4 to 11:3 years in Grade 4 of normal schools on
6 April 1984.

Philippines
All students in Grade 5 classes in regular schools on 1 February 1984.

---

[1] normal or regular indicates that special schools such as
for the handicapped were omitted

Poland
All students in Grade 4 of regular elementary schools with Polish as the official language on 8 May 1984.

Singapore
All students in Primary 5 Normal and Primary 5 Extended Courses on 26 April 1984.

Sweden
All students in Grade 4 in regular schools on 15 May 1983.

United States of America
All students in fifth grade classes of normal schools in the 50 states on 30 April 1986. (It will be noted that the U.S.A. dates of testing are approximately 2 years later than other countries. The U.S.A. undertook data collection in 1983/84 and again in 1986. The 1986 data are presented in this report).

Population 2

Australia
All students of age 14:0 to 14:11 years in normal schools in Years 8, 9 and 10 on 30 September 1983.

Canada (English)
All students in Grade 9 on 1 May 1984.

China
All students in junior secondary Grade 3 in normal schools in Beijing, Tienjin and Taiyuan on 10 May 1985.

England
All students in normal schools in Year 9 in the age range 14:0 to 14:11 on 1 September 1984; that is, of age 13:9 to 14:8 on 1 June 1984.

Finland
All students attending Grade 8 classes in regular Finnish-speaking comprehensive schools on 15 April 1984.

Hong Kong
All students in the second year of secondary education (Grade 8) studying science as part of the Hong Kong Secondary School curriculum on 31 May 1984.

Hungary
All students in the 8th class on 20 May 1983.

Italy
All students of age 14 years in Year 8 (3 Media) and Year 9 (1 Superiore) on 20 April 1983.

Japan
All students in Grade 3 of secondary schools on 15 May 1983.

Korea
All students in the 9th grade on 11 November 1983.

Netherlands
All students in Year 9 of Gymnasium, Atheneum, VWO (general secondary, preuniversity), HAVO (general secondary, senior), MAVO (general secondary, junior), LTO (lower technical vocational), LHNO (lower vocational: administrative and business), LEAO (lower vocational: domestic science), and LAO (lower vocational: agricultural) on 31 May 1984.

Norway
All students of age 15:4 to 16:3 years in Grade 9 in normal schools on 6 April 1984. Norway deliberately chose a year group that was different from the IEA international definition. Grade 9 in Norway is the last grade of full-time compulsory schooling in Norway.

Philippines
All students in Grade 9 of regular schools on 1 February 1984.

Poland
All students in Grade 8 of regular elementary schools with Polish as the official language on 10 May 1984.

Singapore
All students in Secondary 3 on 26 April 1984.

Sweden
All students in Grade 8 of normal comprehensive schools on 15 May 1983.

Thailand
All students in Year 9 in normal schools under the Department of General Education and the Office of the Private Education Commission on 30 December 1983.

United States of America
All students in ninth grade classes of normal schools in the 50 states on 30 April 1986.

Population 3 (Students Studying Science)

Australia
All students in Year 12 on 1 August 1983 who are studying one or more science subjects which meet the prerequisites for entry to tertiary science courses.

Canada (English)
All students in Grade 13 in Ontario and in Grade 12 in all other provinces (except Quebec) and territories in May 1984 who are enrolled in science courses.

England
All students in the second year of an A-level course on 1 March 1984 who are taking science subjects in schools. Students in Colleges of Further Education were omitted. This group represented between 13 and 20 percent of all A level science students depending on science subject. However, from a research study by Keys (1986)[1] it would appear that the College of Further Education students do not have markedly different science achievement from students in ordinary schools.

Finland
All students in the 3rd grade of the Finnish-language Upper Secondary School on 15 November 1983 who are studying one or more of the natural science subjects.

Hong Kong
All students in matriculation classes (forms 6 and 7) on 15 December 1984 who are studying at least one science subject.

Hungary
All students in Grade 4 of academic secondary schools on 20 May 1983 who are studying one or more science subjects for entry into tertiary science courses.

Italy
All students studying any science subject at the final-year secondary level in public or private schools on 20 April 1983. This is Year 12 for Instituti Magistroli and Year 13 for other types of schools.

Japan
All students at the third year upper secondary school level studying Earth Science II and/or Biology II and/or Chemistry II and/or Physics II on 15 November 1983.

Norway
All students in Grade 12 studying one or more science subjects on 20 March, 1984.

Poland
All students in Year 12 in the mathematics/physics and biology/chemistry sections of general education schools on 10 May 1984.

Singapore
All students in Pre-University II classes in Junior Colleges and in Pre-University III classes in Pre-University Centres who are taking one or more science subjects at the GCE A-level on 26 April 1984.

---

[1] Keys, Wendy (1986) A Comparison of A-level science students in schools, sixth-form colleges and colleges of further education. In: Educational Research, Vol.28, No.3, November 1986.

Sweden
All students in Year 12 in the science track (final grade) and in the technology track (Grade 3 out of 4) in the Gymnasium. 30 percent of an age group is in Grade 12 and half of them are in the science or technology tracks. Students in the technology track do not normally study biology. The date of testing was 15 May 1983.

United States of America
All students in the 12th grade of normal schools in the 50 states who are enrolled in second year physics, chemistry and biology courses on 30 April 1986. (In the 1986 testing the U.S.A. did not administer all tests. At Population 3 level, no core test was administered and in the specialist tests a number of items were dropped and the order of items was changed.)

Population 3N (Non-Scientists)

For the countries testing 3N, the target population consisted of all those students not included in Population 3 and the date of testing was the same.

Appendix 2a: Numbers of Schools and Students in Population 1

Country	Schools					Students				
	Target Population	Design Sample	Executed Sample	Achieved Sample	Response Rate	Target Population	Design Sample	Executed Sample	Achieved Sample	Response Rate
Australia	7382	282	220	220	78.01	272891	6363	4259	4259	66.93
Canada (Eng)	7212	314	215	215	68.47	255627	6751	5151	5104	67.20
England	16460	275	181	181	65.82	586467	6048	3748	3748	61.97
Finland	3699	110	106	106	96.36	56274	1857	1631	1600	86.16
Hong Kong a	2460	148	146	146	98.60	92218	5541	5352	5342	96.41
Hungary	3567	100	100	100	100.00	150130	2732	2596	2590	94.80
Italy	31227	205	119	119	58.05	937730	6133	5195	5156	84.07
Japan	20913	223	221	221	99.10	2008068	8000	7925	7924	99.05
Korea	5385	147	146	146	99.32	861868	3528	3489	3489	98.89
Norway	3037	147	91	91	61.90	61358	2420	1329	1305	53.93
Philippines	30216	500	463	463	92.60	1122279	18247	16851	16851	92.35
Poland	14765	199	199	199	100.00	553719	4699	4390	4390	93.42
Singapore	378	251	232	232	92.43	42877	6024	5547	5547	92.08
Sweden	1278	92	64	64	69.57	113063	1960	1518	1449	73.93
U.S.A.	-	140	123	123	87.90	3153653	3665	2861	2822	77.00
Total	147979	3133	2617	2617		10268222	82742	71842	71576	

a In Hong Kong, the primary sampling units were classes rather than schools.

Appendix 2b: Numbers of Schools and Students in Population 2

Country	Schools					Students				
	Target Population	Design Sample	Executed Sample	Achieved Sample	Response Rate	Target Population	Design Sample	Executed Sample	Achieved Sample	Response Rate
Australia	2144	276	233	233	84.42	246114	6624	4917	4917	74.23
Canada (Eng)	2893	316	209	209	66.14	290032	7521	5639	5543	65.80
England	4358	247	147	147	59.51	708039	5928	3118	3118	52.60
Finland	598	93	90	90	96.77	61100	2820	2592	2546	90.28
Hong Konga	2166	133	132	132	99.20	85417	5244	4981	4973	94.83
Hungary	3567	100	99	99	99.00	140968	2704	2515	2515	93.01
Italy	16690	402	291	291	72.39	645158	3644	3249	3228	88.58
Japan	9965	201	199	199	99.00	1819474	8000	7610	7610	95.12
Korea	2116	189	189	189	100.00	837462	4536	4522	4522	99.69
Netherlands	2759	244	224	224	91.80	257595	5856	5065	5025	85.81
Norway	1103	118	77	77	65.25	66931	2429	1424	1420	58.46
Philippines	4117	300	269	269	89.67	573110	12310	10874	10874	88.33
Poland	12607	201	201	201	100.00	488170	4749	4520	4520	95.18
Singapore	249	185	185	185	100.00	39158	4440	4430	4430	99.77
Sweden	948	115	69	69	60.00	120684	2909	1590	1461	50.20
Thailand	2231	103	96	96	93.20	391780	4120	3780	3780	91.75
U.S.A.	-	140	119	119	85.00	2968305	3667	2614	2519	68.70
Total	68511	3363	2819	2819		9739497	87501	73440	73001	

a   In Hong Kong, the primary sampling units were classes rather than schools.

Appendix 3: Number of Schools and Students in Population 3 -All

Country	Schools					Students				
	Target Population	Design Sample	Executed Sample	Achieved Sample	Response Rate	Target Population	Design Sample	Executed Sample	Achieved Sample	Response Rate
Australia		.	165	165	.		.	5057	5057	.
Canada (Eng)			370	370				9925	9452	
England	2736	258	127	127	49.22	149854	9174	3737	3737	40.73
Finland	411	92	86	86	93.48	33022	4033	3775	3638	90.21
Hong Kong (6)	272	163	158	158	96.93	12123	7303	6103	6025	82.50
Hong Kong (7)	194	117	114	114	97.44	6923	4185	3732	3679	87.91
Hungary	250	80	77	77	96.25	19860	2243	2019	2001	89.21
Italy	7309	457	317	317	69.37	392969	9140	6888	6848	74.92
Japan	35180	202	193	193	95.54	1280221	7200	6561	6561	91.12
Norway	214	214	165	165	77.10	9460	3210	1602	1597	49.75
Poland	1121	150	150	150	100.00	40303	3600	3246	3246	90.17
Singapore			16	16				3560	3560	
Sweden	275	175	119	119	68.00	13099	4200	2849	2601	61.93
U.S.A.		.	100	100	.		.	1729	1729	.
Total	47962	1908	2157	2157		1957834	54288	60783	59731	

a   Classes, not schools

Appendix 4: Numbers of Schools and Students in Population 3 - Biology

Country	Schools					Students				
	Target Population	Design Sample	Executed Sample	Achieved Sample	Response Rate	Target Population	Design Sample	Executed Sample	Achieved Sample	Response Rate
Australia	1461	198	164	164	82.83	45483	2276	1631	1631	71.66
Canada (Eng)	1745	286	187	187	65.38	84860	6078	3407	3254	52.71
England			123	123				884	884	
Finland	411	46	43	43	93.48	33022	1975	1707	1652	83.65
Hong Kong (6)			158	158				6021	5960	
Hong Kong (7)			114	114				3654	3614	
Hungary			71	71				304	301	
Italy			12	12				147	147	
Japan	5762a	40	38	38	95.00	194747	1347	1212	1212	89.98
Norway			52	52				277	276	
Poland	518	71	71	71	100.00	19096	1704	764	764	44.84
Singapore	8	8	8	8	100.00	1070	1070	902	902	84.30
Sweden			119	119				1300	1232	
U.S.A.		47	43	43	91.50	348328	858	659	659	76.80
Total	9905	696	1203	1203		726606	15308	22869	22488	

a Classes, not schools.

Appendix 5: Numbers of Schools and Students in Population 3 - Chemistry

Country	Schools					Students				
	Target Population	Design Sample	Executed Sample	Achieved Sample	Response Rate	Target Population	Design Sample	Executed Sample	Achieved Sample	Response Rate
Australia	1461	199	164	164	82.41	30530	1520	1177	1177	77.43
Canada (Eng)	1662	297	179	179	60.27	74760	5732	3108	2923	51.40
England			123	123				892	892	
Finland	411	46	44	44	95.65	13209	1167	1001	971	83.20
Hong Kong (6)			158	158				6081	6018	
Hong Kong (7)			114	114				3711	3670	
Hungary			56	56				146	143	
Italy			24	24				217	217	
Japan	7209a	43	43	43	100.00	268176	1584	1468	1468	92.68
Norway			46	46				284	283	
Poland	518	71	71	71	100.00	19096	1704	765	765	44.89
Singapore	8	8	8	8	100.00	1270	1270	945	945	74.41
Sweden			119	119				1261	1172	
U.S.A.		47	40	40	76.10	107470	771	537	537	69.60
Total	11269	711	1189	1189		514511	13784	21593	21181	

a   Classes, not schools.

Appendix 6: Number of schools and students in Population 3 - Physics

Country	Schools					Students				
	Target Population	Design Sample	Executed Sample	Achieved Sample	Response Rate	Target Population	Design Sample	Executed Sample	Achieved Sample	Response Rate
Australia	1461	198	163	163	82.32	26679	1420	1073	1073	75.56
Canada (Eng)	1544	281	181	181	64.41	57560	5221	2890	2766	53.60
England			125	125				917	917	
Finland	411	46	42	42	91.30	11558	977	837	810	82.91
Hong Kong (6)			158	158				6077	6025	
Hong Kong (7)			114	114				3719	3679	
Hungary			75	75				400	398	
Italy			120	120				1773	1766	
Japan	5521a	39	36	36	92.31	190490	1332	1187	1187	89.11
Norway			55	55	100.00			444	443	
Poland	603	79	79	79	100.00	21207	1896	1716	1716	90.51
Singapore	8	8	8	8		1308	1308	1071	1071	81.88
Sweden			119	119				1231	1156	
U.S.A.		46	35	35	76.10	28722	762	485	485	63.65
Total	9548	697	1310	1310		337524	11480	23820	23482	

a Classes, not schools

Appendix 7: Numbers of Schools and Students in Population 3 - Non-Scientists

Country	Schools					Students				
	Target Population	Design Sample	Executed Sample	Achieved Sample	Response Rate	Target Population	Design Sample	Executed Sample	Achieved Sample	Response Rate
Australia	1461	200	120	120	60.00	98432	1907	995	995	52.18
Canada(Eng)	.	.	81	81	.	.	.	520	509	.
England	.	.	126	126	.	.	.	1004	1004	.
Hungary	.	.	77	77	.	.	.	1046	1036	.
Italy	.	.	119	119	.	.	.	2476	2455	.
Japan	15556a	.	60	60	96.77	589561	2347	2229	2229	94.97
Norway	222	111	78	78	70.27	13135	1665	597	595	35.74
Singapore	.	.	9	9	.	.	405	297	297	73.33
Total	17239	373	670	670		701128	6324	9164	9120	

a   Classes, not schools.

8 7

Appendix 8:   Calculation of Test Validity Indices

An extremely important question in an international study is that of the validity of the tests. Three indices have been developed to address this problem. The first question is "To what extent does the test cover the planned or intended curriculum or syllabus?" It is clear that a school system where only ten percent of its curriculum is covered by a test cannot be expected to have a level of performance comparable with that of another system where ninety percent of the curriculum is covered. An index called the curriculum relevance index has been calculated which is a measure of the extent to which the curriculum of a particular system is covered by the items in the test. The second question is "To what extent are the items in the test relevant to the curriculum of a country?" In order to measure this, a test relevance index was developed. However, even when all the items in a test are relevant to a system's curriculum, they may only cover a small percentage of that system's curriculum. The curriculum relevance index allows for this aspect while the test relevance index does not.

The final question is "To what extent does the curriculum of a system match the total curriculum of all systems together?" The measure developed is a curriculum coverage index.

What follows describes how the indices were calculated.

Curriculum Ratings

In 1970, there had been a first science study so that one constraint in designing the second study was that there should be as much in common as possible between the first and second studies to enable sound comparisons to be made across occasions and across countries with regard to achievement in science. The first study in 1970 had used 53 content areas to define the content of the science curriculum in the form of a curriculum grid. The 53 content areas were the result of content analyses carried out in collaboration with the participating countries in the first study (Comber and Keeves, 1973). In 1980, various content areas were added to the curriculum grid to give a total of 57 content areas. This grid was sent out to all the participating countries. Each National Research Coordinator was responsible for obtaining ratings for each topic according to a four point scale:

3 =   content area is given major emphasis in the curriculum for all students;

2 =   content area is given major emphasis in the curriculum for some students; or minor emphasis for all students;

1 =   content area is given minor emphasis in the curriculum for some students; and

0 =   content area is not included in curriculum

Rosier (1987, p. 112)[1] has stated:

Where the curriculum was prescribed centrally in a system, it was relatively easy for the NRCs (National Research Coordinators) in association with their National Committees to assign the ratings mainly on the basis of an examination of published curriculum statements but also on the basis of textbooks and examination papers. In systems where the curricula were more decentralized (as in Australia), it was necessary to conduct a survey of schools or persons with responsibility for the science curriculum to determine the curriculum being used and then to collate the responses to give an overall rating for the system.

Table 2 in Chapter 1 has provided for each population tested a list of these content areas together with the extent of coverage across all 24 countries, and the number of items included in each of the tests that were administered.

The particular values assigned by a country for each cell in the curriculum grid are referred to as the curriculum emphasis scores (see Table A 8.1)

---

1   Rosier, M.J., (1987) The second international science study Comparative Education Review, 31(1): 106-128

Calculation of Curriculum Indices

As a first step in the calculation of these three indices the content grid emphasis scale was collapsed from a 4-point to a 3-point scale, because of substantial difficulties which were encountered in some countries and were evident in the responses from those countries concerned with whether major emphasis applied to all or to only some students. Thus, all ratings of value 3 were reduced down to 2, while the 1 and 0 ratings were left unchanged. The maximum coverage score for each content area in the grid was thus 2, and the total of these maximum coverage scores was used in the calculation of the curriculum coverage index. In order to explain exactly how the three indices were calculated the following hypothetical table is presented.

Table A 8.1: Hypothetical table for calculating validity indices

(1) Content Area	(2) Maximum Coverage Score	(3) Curriculum Emphasis Score	(4) Number of Items	(5) Item Relevance Score	(6) Max. Item Relevance Score	(7) Cell Relevance Score
A	2	2	4	(4x2)	(4x2)	2
B	2	1	3	(3x1)	(3x2)	1
C	2	2	2	(2x2)	(2x2)	2
D	2	0	1	0	2	-1
E	2	1	0	0	0	0
F	2	0	0	0	0	0
Total	12	6	10	15	20	4

Calculation of Indices:

Curriculum Coverage Index	=	6/12	= 0.50
Test Relevance Index	=	15/20	= 0.75
Curriculum Relevance Index	=	4/6	= 0.67

Curriculum Relevance Index

Two aspects of curriculum relevance have to be taken into consideration. First, it is necessary to consider whether the items in the test are associated with a content area in which there is major, minor or no emphasis in the particular country. Secondly, it is necessary to consider whether for each curriculum content area given an emphasis rating in the curriculum of a country there is an appropriate item in the test. Where the curriculum emphasis score is 2 and there are items in the test a cell relevance score of 2 is assigned. Where there is a cell with only a minor curriculum emphasis and there are items in the test a score of 1 is assigned. However, if there is a cell with no emphasis on the content area and there are items in the test a score of -1 is assigned. Where there is a cell with major or minor emphasis, but there is no item in the test a cell relevance score of 0 is assigned. The cell relevance scores are entered in column 7 of Table 3 and the total cell relevance score is calculated. It will be seen that the cell relevance scores reflect the curriculum emphasis of a particular country and the curriculum relevance index is calculated by dividing the total cell relevance score (4) by the curriculum emphasis score (6) to give the value 0.67.

Test Relevance Index

In column 4 of Table 3 the number of items in the test is recorded. If an item lies in a content area in which for a particular country the curriculum emphasis score is 2, then a value of 2 is assigned as an item relevance score, and if the curriculum emphasis score is 1, a value of 1 is assigned. For each content area the number of assigned item relevance scores is equal to the number of items associated with that content area, and the sum of the assigned item relevance scores is entered in column 5. The maximum item relevance score for a content area is entered in column 6 and is obtained by multiplying the number of items associated with that content area by 2 which is the maximum coverage score for the cell. In Table 3 the sum of the item relevance scores and the maximum item relevance scores are calculated and are entered at the foot of columns 5 and 6 respectively. The test relevance index gives the extent to which the items in the test are relevant to the curriculum of a country and is calculated by dividing the total item relevance score (15) by the maximum total item relevance score (20) to give the value 0.75.

Curriculum Coverage Index

From each country the curriculum emphasis score on the three point scale was entered in column 3, alongside the maximum coverage score which was entered in column 2. In Table 3, for the fictitious data, the total maximum coverage score is 12 and the total curriculum emphasis score is 6. The extent to which the curriculum of a country covers the complete grid is given by the curriculum coverage index which is calculated by dividing the total curriculum emphasis score (6) by the total maximum coverage score (12) to obtain the value of the index of (0.50).

Tables A 8.2 and A 8.3 present the three indices for each population for the core test plus rotated items and core test respectively. It can be seen that the curriculum relevance indices for the core test plus rotated items are higher than for the core test only. This is to be expected since the totality of the tests covers more than just the core test only. Given that the correlations among core test and rotated tests are high, the comments here refer to Table A 8.2. With the exception of Japan, Singapore, and Sweden the tests are covering about 60 percent of the national curricula at Population 1 level. At Population 2 level, with the exception of Singapore, the coverage is between 54 and 70 percent. For the biology specialist test the coverage is 76 percent for most countries and for chemistry and physics the coverage is 80 to 90 percent.

Table A 8.2: Test validities: curriculum relevance index; test relevance index; curriculum coverage index (core plus rotated tests)

	Australia	Canada (English)	England	Finland	Hong Kong	Hungary	Italy	Japan	Korea	Netherlands	Norway	Philippines	Poland	Singapore	Sweden	Thailand	U.S.A.
Population 1																	
Curriculum Relevance Index	.67	.49	.66	.61	n.a.	.58	.68	.38	n.a.	-	n.a.	.66	.64	.25	.43	-	.62
Test Relevance Index	.86	.85	.96	.79	n.a.	.73	.79	.55	n.a.	-	n.a.	.91	.79	.50	.56	-	.86
Curriculum Coverage Index	.44	.54	.61	.44	n.a.	.37	.37	.22	n.a.	-	n.a.	.53	.44	.28	.30	-	.57
Population 2																	
Curriculum Relevance Index	.70	.49	.64	.61	.56	.68	.61	.57	n.a.	.61	n.a.	.51	.68	.40	.54	.56	.61
Test Relevance Index	.93	.67	.94	.86	.91	.99	.92	.86	n.a.	.94	n.a.	.73	.96	.61	.91	.89	.94
Curriculum Coverage Index	.66	.58	.76	.82	.59	.78	.66	.62	n.a.	.79	n.a.	.64	.72	.36	.78	.65	.81
Population 3																	
Biology																	
Curriculum Relevance Index	.76	.76	.84	.76	.76	.76	.76	.76	.76	-	n.a.	-	.76	.76	.76	.68	.76
Test Relevance Index	1.00	1.00	1.00	1.00	1.00	1.00	1.00	1.00	1.00	-	n.a.	-	1.00	1.00	1.00	1.00	1.00
Curriculum Coverage Index	1.00	1.00	.91	1.00	1.00	1.00	1.00	1.00	1.00	-	n.a.	-	1.00	1.00	1.00	.88	1.00
Chemistry																	
Curriculum Relevance Index	.96	.18	.96	.93	.96	.88	.93	.93	.93	-	n.a.	-	.96	1.00	.96	1.00	.93
Test Relevance Index	.96	.31	.96	.97	.97	.92	.96	1.00	1.00	-	n.a.	-	.91	1.00	.88	1.00	1.00
Curriculum Coverage Index	.90	.37	.90	.97	.87	.83	.93	1.00	1.00	-	n.a.	-	.80	.93	.87	.93	1.00
Physics																	
Curriculum Relevance Index	.94	.86	.89	.87	.97	.89	.88	.89	.97	-	n.a.	-	.92	.85	.89	.85	.85
Test Relevance Index	.96	.89	1.00	1.00	.95	.99	.95	1.00	.95	-	n.a.	-	1.00	1.00	.97	1.00	1.00
Curriculum Coverage Index	.83	.88	.95	.98	.80	.95	.85	.95	.75	-	n.a.	-	.93	1.00	.93	1.00	1.00

n.a. = not available
- = this population not tested

Table A 8.3: Test validities: curriculum relevance index; test relevance index; curriculum coverage index (core plus rotated tests)

	Australia	Canada (English)	England	Finland	Hong Kong	Hungary	Italy	Japan	Korea	Netherlands	Norway	Philippines	Poland	Singapore	Sweden	Thailand	U.S.A.
Population 1																	
Curriculum Relevance Index	.45	.33	.37	.37	n.a.	.28	.48	.35	n.a.	-	n.a.	.45	.51	.28	.37	-	.45
Test Relevance Index	.88	.85	.94	.81	n.a.	.77	.83	.67	n.a.	-	n.a.	.94	.83	.50	.65	-	.94
Curriculum Coverage Index	.44	.54	.61	.44	n.a.	.37	.37	.22	n.a.	-	n.a.	.53	.41	.29	.30	-	.57
Population 2																	
Curriculum Relevance Index	.44	.27	.34	.34	.49	.37	.42	.39	n.a.	.37	n.a.	.34	.39	.36	.32	.36	.34
Test Relevance Index	.98	.76	.93	.98	.59	.98	.92	.90	n.a.	1.00	n.a.	.78	.98	.67	.90	.88	.95
Curriculum Coverage Index	.66	.58	.76	.82	.97	.78	.66	.62	n.a.	.79	n.a.	.64	.72	.36	.78	.65	.81
Population 3																	
Biology																	
Curriculum Relevance Index	.65	.65	.71	.65	.65	.65	.65	.65	.65	n.a.	n.a.	-	.65	.65	.65	.59	.65
Test Relevance Index	1.00	1.00	1.00	1.00	1.00	1.00	1.00	1.00	1.00	n.a.	n.a.	-	1.00	1.00	1.00	1.00	1.00
Curriculum Coverage Index	1.00	1.00	.91	1.00	1.00	1.00	1.00	1.00	1.00	n.a.	n.a.	-	1.00	1.00	1.00	.88	1.00
Chemistry																	
Curriculum Relevance Index	.93	.09	.93	.86	1.00	.92	.89	.87	.87	n.a.	n.a.	-	1.00	.93	.88	.93	.87
Test Relevance Index	.97	.37	.97	.97	1.00	.93	.97	1.00	1.00	n.a.	n.a.	-	.93	1.00	.88	1.00	1.00
Curriculum Coverage Index	.90	.37	.90	.97	.87	.83	.93	1.00	1.00	n.a.	n.a.	-	.80	.93	.87	.93	1.00
Physics																	
Curriculum Relevance Index	.94	.86	.89	.87	.97	.89	.88	.89	.97	n.a.	n.a.	-	.92	.85	.89	.85	.85
Test Relevance Index	.95	.92	1.00	1.00	.93	.98	.93	1.00	.93	n.a.	n.a.	-	1.00	1.00	.98	1.00	1.00
Curriculum Coverage Index	.83	.88	.95	.98	.80	.95	.85	.95	.75	n.a.	n.a.	-	.93	1.00	.93	1.00	1.00

n.a. = not available
- = this population not tested

For the test relevance or appropriateness to the curriculum the indices are higher and typically 80 and 90 percent but again with the exception of Japan, Singapore and Sweden at Population 1 level and the Philippines and Singapore at Population 2 level. For the Population 3 specialist tests the relevance is very high.

For the curriculum coverage indices, the extent to which the national curriculum covers the universal curriculum, there are interesting differences. For Population 1 the range is from 22 to 61 percent, for Population 2 from 36 to 82 percent, but for the specialist tests the differences are small ranging from 75 to 100 percent. Table A 8.3 presents a similar profile to the indices in Table A 8.2.

In Chapters 1 and 5 only the curriculum indices for the core tests and the Population 3 tests for students studying science are considered because it is the achievement test scores for these tests that are being reported in this volume. There are two uses for these indices. First, they provide evidence of the relative meaningfulness of the tests which are employed in this study in the different countries that have taken part. Secondly, they provide measures of the intended curriculum, with respect to the tests which were used in this study, that can be employed to relate the intended curriculum to achievement outcomes. The first aspect is considered in Chapter 1 and the second aspect in Chapter 5.

It is hoped that in the further reporting of this study, the data from the rotated tests will be analyzed. The total test scores consisting of the core plus rotated tests clearly have greater validity as assessed by these three indices than the core and specialist tests by themselves, and because of their greater internal consistency, measured by the reliability coefficient, and their greater strength, measured by the validity coefficients, their use in all further analyses is warranted.

Appendix 9: Test Reliabilities and Standard Errors of Sampling

Country	Population 1 Core Test (24 items)		Population 2 Core Test (30 items)		Population 3a Core Test (30 items)		3B (30 items)		3C (30 items)		3P (30 items)		3N (30 items)	
	KR-20	SES	KR-20	SES	KR-20	SES	KR-20	SES	KR-20	SES	KR-20	SES	KR-20	SES
Australia	.761	.175	.774	.189	.831	.168	.630b	.151	.801	.305	.707	.212	.717	.231
Canada (Eng)	.736	.130	.754	.166	.828	.144	.656	.195	.722	.308	.664c	.200	.712	.341
England	.755	.168	.771	.219	.833	.151	.624	.236	.798	.285	.713	.200	.645	.169
Finland	.713	.150	.703	.129	.790	.138	.780	.160	.832	.259	.850	.274	-	-
Hong Kongd	.716	.195	.716	.249	.642	.189	.686	.271	.779	.401	.706	.341	-	-
Hungary	.776	.231	.796	.255	.801	.298	.663	.371	.818	.697	.805	.501	.665	.281
Italy	.793	.255	.780	.276	.817	.230	.663	1.046	.890	1.453	.673	.247	.711	.273
Japan	.715	.068	.800	.088	.867	.340	.690	.475	.873	.860	.785	.584	.648	.291
Korea	.738	.157	.745	.145	-	-	-	-	-	-	-	-	-	-
Netherlands	-	-	.794	.257	-	-	-	-	-	-	-	-	-	-
Norway	.712	.300	.753	.155	.837	.229	.693	.334	.769	.370	.731	.332	.644	.236
Philippines	.771	.159	.735	.200	-	-	-	-	-	-	-	-	-	-
Poland	.768	.159	.794	.223	.779	.216	.591	.285	.780	.444	.777	.420	.588	-
Singapore	.737	.177	.765	.280	.783	-	.630	-	.794	-	.637	-	-	-
Sweden	.703	.156	.782	.223	.752	.149	.725	.230	.761	.234	.694	.178	-	-
Thailand	-	-	.664	.221	-	-	-	-	-	-	-	-	-	-
U.S.A.e	.770	.179	.770	.272	-	-	.669	.405	.765	.670	.695	.526	-	-

a Standard errors of sampling for all Population 3 score means are given in raw score units.

b Australia had only 29 items in the Biology Test because one item in the test had a typing error in it which rendered it invalid.

c Canada (Eng) requested that one item be dropped from the Physics Test because one phrase had been omitted in the printed version of the item.

d At Population 3 Hong Kong tested Forms 6 and 7 separately. The figures in the above table are for form 6.
For Form 7 they are:

3A		3B		3C		3P	
KR-20	SES	KR-20	SES	KR-20	SES	KR-20	SES
.610	.185	.764	.277	.843	.430	.727	.375

e In the United States 5 items were dropped from both the Biology and Chemistry Tests and 4 items from the Physics Test.

Appendix 10

The Calculation of Growth Scores

The growth scores between the three levels can be calculated through a procedure of equating that employs not only performance on the anchor items reported on Table 14, but also performance on the core tests which were taken by the students at the three grade levels, which have been reported in Tables 9 to 12. The anchor items are used to locate the score levels for each population on a common scale using standard score equating procedures, and performance on the core tests is used to locate the mean score for each country above or below the population score level. So that subsequently equating could be undertaken across occasions from 1970-71 to 1983-84, only those countries that tested on both occasions were used in the equating across population levels reported in this Appendix. For a more detailed account of the equating procedures employed, the report of the First International Science Study (Comber and Keeves, 1973, p. 167-168) should be consulted. It should be noted however, that assessing growth in this way does not provide a measure of absolute gain in performance from one grade level to the next as is provided by performance on the anchor items. The scaling procedures employed permitted the relative performance of each country at the different levels to be recorded on a single common scale in such a manner that differences between countries and population levels could be examined. The use of performance on the core test, rather than a limited subset of items, the anchor items, from the core tests provided a more effective sampling of student achievement at each of the population levels. The core test provides scores which are above or below the reference levels on the common scale that was determined by the anchor items.

In Table A 10, the standard scores for the students at each population level for each country are recorded for this common scale. From these standard scores, growth scores have been calculated between Populations

Table A 10: Standard Scores and Growth Scores for All Countries on a Common Scale

Country	Standard Scores				Growth Scores			
	Pop. 1	Pop. 2	Pop. 3S	Pop. 3N	1-->2	1-->3S	2-->3S	2-->3N
Australia	-1.36	-0.10	1.21	0.02	1.26	2.57	1.31	0.12
Canada	-1.13	0.06	0.97	-0.43	1.18	2.10	0.92	-0.37
England	-1.67	-0.33	1.91	0.65	1.34	3.58	2.24	0.98
Finland	-0.71	0.04	1.07	-	0.75	1.77	1.02	-
Hong Kong (Form 6)	-1.81	-0.40	1.76	-	1.40	3.57	2.16	-
Hong Kong (Form 7)	-	-0.40	2.00	-	-	3.81	2.40	-
Hungary	-0.94	0.69	1.89	1.47	1.63	2.82	1.19	0.78
Italy	-1.22	-0.33	0.63	-0.30	0.89	1.85	0.96	0.03
Japan	-0.67	0.39	1.58	0.59	1.06	2.25	1.19	0.20
Korea	-0.69	-0.06	-	-	0.64	-	-	-
Norway	-1.41	-0.08	1.30	-0.12	1.33	2.70	1.38	-0.04
Philippines	-2.26	-1.42	-	-	0.84	-	-	-
Poland	-1.61	-0.04	1.28	-	1.57	2.89	1.32	-
Singapore	-1.79	-0.38	1.65	0.10	1.41	3.26	2.03	0.48
Sweden	-0.88	0.22	1.59	-	0.90	2.47	1.57	-
U.S.A.	-1.28	-0.39	-	-	0.90	-	-	-

1 and 2, Populations 1 and 3 ,and Populations 2 and 3. In further reporting for this study, attempts will be made to employ latent trait procedures to test whether the anchor items which are used to locate the necessary fixed points on the common scale are consistent with the assumption of a single underlying trait. In addition, attempts will be made to assess the relative position of a country above or below the fixed points for each population using not only performance on the core test, but also performance on the rotated tests that were administered.

Appendix 11: Rank orders of countries in 1970 and the mid-1980s

	Pop 1		Pop 2		Pop 3	
	1970	1980s	1970	1980s	1970	1980s
Australia	-	9	3	10	3	9
Belgium (Fl)	3	-	10	-	11	-
Belgium (Fr)	12	-	14	-	13	-
Canada (Eng)	-	6	-	4	-	10
Chile	14	-	15	-	17	-
England	8	12	9	11	5.5	3
Fed. Rep. Germany	10	-	5	-	2	-
Finland	5	3	7	5	8	9
France	-	-	-	-	10	-
Hong Kong	-	13	-	16	-	1
Hungary	6	5	2	1	7	2
India	15	-	17	-	18	-
Iran	16	-	16	-	16	-
Italy	7	7	11	11	12	12
Japan	1	1	1	2	-	7
Korea	-	1	-	7	-	n.a.
Netherlands	9	-	12	3	4	-
New Zealand	-	-	4	-	1	-
Norway	-	10	-	9	-	11
Philippines	-	15	-	17	-	-
Poland	-	11	-	7	-	6
Scotland	11	-	8	-	5.5	-
Singapore	-	13	-	14	-	5
Sweden	2	4	6	6	9	4
Thailand	13	-	13	14	15	n.a.
U.S.A.	4	8	7	14	14	-

Appendix 12

<u>Some Sample Test Items (Populations 1, 2 and 3)</u>

This appendix presents the item analyses for selected items from the science achievement tests.

In each case, the item is presented and this is followed by the item analyses for the participating countries. In all item statistic tables, N is the number of students in the sample. A, B, C, D, and E represent the options and the figures beneath are the percentages of students in each country selecting each option. The option with the star is the correct response. In the first item it is A. Therefore, B, C, D, and E represent wrong answers. The title 'None' represents the percentage of the sample not responding to the item.

Under the item analyses, comments have been made on performance on the item.

The tests were prepared with equal consideration given to the curricula of all countries engaged in the study at the time the tests were developed. The estimates of the validity indices for the different countries reported in Chapter 1 and Appendix 8 indicate the appropriateness of the tests in the different countries. In general, the test relevance indices, which signfy the extent to which the test items are in the different curricula, are high. Nevertheless, there are cases where the content associated with an item is not taught to all students in a country or not taught at all. In the comments that follow, in fairness to the countries involved, we have drawn attention to situations where this seems to be occurring. It is consequently necessary for curriculum developers to examine their syllabus, the time given to the study of science and their objectives for science education to assess the appropriateness of the comments provided for each item and the need for possible change in their science courses.

Population 1

The Population 1 core test included items classified as earth science, biology, chemistry or physics items. Seven items are presented below. The first is an earth science item which is then followed by two items from each of the other areas.

The sun is the only body in our solar system that gives off large amounts of light and heat. Why can we see the moon ?

SASC—H

A   It is reflecting light from the Sun.
B   It is without an atmosphere.
C   It is a star.
D   It is the biggest object in the solar system.
E   It is nearer the Earth than the Sun.

Country	N	A*	B	C	D	E	None
Australia	4259	48.83	3.32	9.56	7.57	29.65	1.07
Canada(Eng)	5104	61.26	3.25	10.10	3.78	20.43	1.18
England	3748	47.71	4.13	11.63	4.85	30.53	1.16
Finland	1600	68.16	4.46	5.67	3.09	18.33	0.29
Hong Kong	5342	43.89	5.94	9.79	12.65	26.99	0.75
Hungary	2590	67.70	4.17	4.49	7.24	14.84	1.56
Italy	5156	55.58	5.25	10.52	2.91	22.96	2.79
Japan	7924	66.25	4.10	4.16	4.48	20.80	0.22
Korea	3489	58.66	3.58	9.24	3.08	25.32	0.12
Norway	1305	55.62	6.53	7.89	5.22	23.07	1.67
Philippines	16851	61.43	4.94	6.71	12.10	14.33	0.48
Poland	4390	41.34	7.20	8.70	14.69	24.56	3.52
Singapore	5547	56.27	7.59	3.91	7.61	24.42	0.20
Sweden	1449	70.15	1.29	6.72	3.74	16.95	1.16
U.S.A.	2822	65.57	2.40	8.58	4.69	18.59	0.18

The item is of average difficulty and the percent in each country selecting the correct response (A) ranged from 41 percent in Poland to 70 percent in Sweden.
The popularity of distractor E in all coutries suggests that, while most students learn something about the solar system at school or in informal ways, the fact that the Sun is the source of light that enables us to see the moon is not explained to all primary school students. Moreover, the fact that approximately 10 percent of students in Australia, Canada (English), England and Italy consider the moon to be a star, would seem to indicate that some students of 10 years of age in those countries have not been taught about the solar system.

Milk kept in a refrigerator does not go sour. Why ?

A   The cold changes the water of the milk into ice.
B   The cold separares the cream.
C   The cold slows down the action of bacteria.
D   The cold keeps flies away.
E   The cold causes a skin to form on the surface of the milk.

Country	N	A	B	C*	D	E	None
Australia	4259	6.98	19.96	46.51	3.65	21.61	1.29
Canada(Eng)	5104	5.83	17.71	60.17	1.84	13.37	1.09
England	3748	8.01	22.84	40.63	4.54	22.29	1.68
Finland	1600	7.07	15.93	53.87	5.27	17.62	0.23
Hong Kong	5342	12.26	26.04	30.44	4.60	26.30	0.37
Hungary	2590	17.41	8.57	52.69	2.99	16.95	1.40
Italy	5156	4.04	12.06	55.35	3.82	22.30	2.44
Japan	7924	5.71	11.11	51.64	2.21	28.52	0.80
Korea	3489	13.52	5.02	55.06	1.85	24.50	0.06
Norway	1305	6.88	32.94	29.20	6.54	21.90	2.55
Philippines	16851	29.54	15.91	29.01	6.85	18.26	0.41
Poland.	4390	19.35	19.67	36.62	4.35	16.35	3.66
Singapore	5547	17.84	19.42	32.78	6.12	23.35	0.50
Sweden	1449	5.42	28.68	52.76	2.47	8.30	2.37
U.S.A.	2822	6.71	14.70	63.31	2.90	11.76	0.62

Again, this item is of average difficulty for most countries (the correct answer being C) but it is of interest to note that Hong Kong, the Philippines, and Singapore have relatively low scores. These are all Asian countries where it is relatively hot and humid all the time, and the need to keep milk from turning sour would seem to be of some consequence in these countries. Option E is the most popular distractor in most countries, although a surprisingly small number of Swedish students selected that option. The marked popularity of option B in Norway (33%) and Sweden (29%) but not in Finland (16%) is of interest and warrants examination by curriculum specialists in these countries.

Flowers cannot usually produce seeds unless

A   they are visited by insects.
B   they appear in the summer.
C   they are on plants growing in good soil.
D   they produce nectar.
E   suitable pollen is placed on their stigmas.

Country	N	A	B	C	D	E*	None
Australia	4259	14.39	11.68	36.02	11.92	24.70	1.29
Canada(Eng)	5104	15.32	11.71	34.35	13.93	22.78	1.91
England	3748	21.75	19.54	28.43	13.10	15.17	2.01
Finland	1600	32.28	8.40	6.53	4.35	47.95	0.48

Country	N	A	B*	C	D	E	None
Hong Kong	5342	29.86	9.28	22.45	9.92	27.95	0.54
Hungary	2590	3.55	5.79	38.81	3.36	47.27	1.22
Italy	5156	14.71	10.27	28.35	6.31	37.42	2.94
Japan	7924	10.88	10.82	20.91	5.71	51.16	0.52
Korea	3489	16.17	5.87	12.26	5.26	60.44	0.00
Norway	1305	28.89	11.65	44.86	1.14	12.17	1.29
Philippines	16851	27.49	8.95	18.98	15.03	29.09	0.46
Poland	4390	36.98	17.66	14.20	6.03	21.70	3.43
Singapore	5547	24.27	10.08	24.20	16.42	24.73	0.30
Sweden	1449	32.29	8.48	7.92	14.39	33.96	2.96
U.S.A.	2822	10.80	9.40	40.10	10.47	28.39	0.85

This item is a relatively difficult item since the correct responses (E) range from 12 percent in Norway to 60 percent in Korea. Options A and C are particularly strong distractors.

It would seem that the content of this item is unfamiliar to the 10-year old students in most countries. However, the relatively high level of correct response to this item in Korea (60%), Japan (51%), Finland (48%) and Hungary (47%) suggests that only in these countries is reproduction in plants taught at this grade level.

Three candles, which are exactly the same, are placed in different boxes as shown in the diagram. Each candle is lit at the same time.

Large closed box

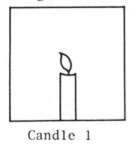

Candle 1

Small closed box   Open box

Candle 2   Candle 3

In what order are the candle flames most likely to go out?

A   1, 2, 3
B   2, 1, 3
C   2, 3, 1
D   1, 3, 2
E   3, 2, 1

Country	N	A	B*	C	D	E	None
Australia	4259	11.58	40.32	7.46	4.67	34.69	1.27
Canada(Eng)	5104	9.92	53.64	6.76	3.74	25.54	0.40
England	3748	11.84	38.13	6.10	4.90	37.08	1.94
Finland	1600	8.12	67.16	5.91	3.37	15.07	0.38
Hong Kong	5342	10.53	40.13	7.61	6.63	34.52	0.58
Hungary	2590	12.34	69.74	3.78	1.92	10.42	1.79
Italy	5156	10.02	62.05	5.70	1.66	16.70	3.88
Japan	7924	6.13	53.17	7.40	7.87	25.08	0.35
Korea	3489	8.48	53.86	5.42	4.39	27.79	0.06
Norway	1305	11.10	63.70	5.74	3.40	13.29	2.76
Philippines	16851	16.23	25.09	11.52	7.83	38.89	0.43
Poland	4390	12.94	56.05	6.97	2.55	17.87	3.61
Singapore	5547	8.68	26.67	7.07	7.95	49.28	0.34
Sweden	1449	6.87	71.19	5.46	2.35	12.01	2.11
U.S.A.	2822	11.67	44.04	6.90	5.35	31.69	0.34

The correct response to this item is option B. However, to answer this item correctly more than knowledge and understanding is required and some 10-year old students may not have the cognitive skills necessary to provide a correct answer to such an item.

Again, this is an item of average difficulty but it is surprising how many children selected the wrong answer with the open box candle going out first. Do 10-year old children envisage that a breeze is likely to blow out Candle 3, where it is set in the open box. However, the low level of correct response in England (38%), Singapore (27%) and the Philippines (25%) suggests that in these countries the burning of a candle in air is not part of the learning experiences of most students in science classes in these countries.

Paint applied to an iron surface prevents the iron from rusting.Which <u>one</u> of the following provides the best reason?

A   It prevents nitrogen from coming in contact with the iron.
B   It reacts chemically with the iron.
C   It prevents carbon dioxide from coming in contact with the iron.
D   It makes the surface of the iron smoother.
E   It prevents oxygen and moisture from coming in contact with the iron.

Country	N	A	B	C	D	E*	None
Australia	4259	10.08	9.16	13.66	17.17	45.77	4.17
Canada(Eng)	5104	9.86	10.04	14.05	13.83	48.29	3.93
England	3748	10.95	12.02	18.10	15.77	35.76	7.40
Finland	1600	7.48	4.56	6.43	10.56	70.16	0.81
Hong Kong	5342	9.67	14.51	12.65	17.10	45.35	0.71
Hungary	2590	9.77	9.15	7.92	7.21	63.67	2.28
Italy	5156	11.47	8.01	6.67	7.61	61.93	4.31
Japan	7924	13.66	6.24	19.85	11.73	47.75	0.77
Korea	3489	14.06	6.45	8.70	9.15	61.52	0.12
Norway	1305	12.90	8.50	8.74	9.62	55.71	4.53
Philippines	16851	11.66	16.35	23.14	13.08	34.94	0.83
Poland	4390	13.84	4.81	12.66	7.51	54.74	6.45
Singapore	5547	15.89	13.94	17.31	20.43	32.09	0.33
Sweden	1449	7.43	11.86	7.67	4.43	63.87	4.74
U.S.A.	2822	13.03	10.52	16.09	13.59	45.63	1.14

The correct response to this item is E and, again, it is an item of average difficulty ranging from 32 percent in Singapore to 70 percent in Finland and 64 percent in Hungary. Although the chemical principles underlying rusting are rather more complex than those involved in the previous question on burning, the level of correct responses is similar across countries. Once again the students in England (36%), Philippines (35%) and Singapore (32%)were much less successful on this item than students in other countries. This suggests that consideration of rusting is not taught in the science courses of these countries at the 10-year old level.

Why are two holes often punched in a can of juice before pouring it?

A   to let the juice pour out of the can more slowly
B   to let the air go into one hole while the juice pours out of the other
C   to let the air get into the can before the juice is poured
D   to let the juice pour out of the can more quietly
E   to watch how the juice is pouring out

Country	N	A	B*	C	D	E	None
Australia	4259	12.32	52.70	21.31	7.98	3.51	2.19
Canada(Eng)	5104	7.57	63.86	17.13	9.17	1.37	0.90
England	3748	19.59	37.74	27.49	7.73	4.93	2.52
Finland	1600	7.26	47.35	35.78	5.66	3.51	0.43
Hong Kong	5342	15.91	51.33	15.56	10.71	5.80	0.69

Country	N	A	B*	C	D	E	None
Hungary	2590	6.44	39.32	10.22	32.72	9.48	1.82
Italy	5156	23.22	49.20	12.14	4.94	6.49	4.02
Japan	7924	11.61	67.30	11.35	5.03	3.97	0.75
Korea	3489	7.94	66.54	9.28	11.08	5.13	0.03
Norway	1305	12.66	38.81	14.50	13.46	17.75	2.81
Philippines	16851	13.65	42.60	18.53	11.79	12.74	0.69
Poland	4390	9.85	47.54	12.35	16.65	7.10	6.51
Singapore	5547	14.45	28.51	18.04	31.63	6.86	0.51
Sweden	1449	22.57	31.83	31.27	4.82	6.34	3.16
U.S.A.	2822	15.97	57.28	12.82	10.95	2.24	0.75

The correct response to this physics item is B with a range from 29 percent in Singapore to 67 percent in Japan. A and C were strong distractors and the high percentage of Hungarians (33%) and Singaporeans (32%) choosing option D suggests that there may be differences between countries in the extent to which metal cans are used as containers for fruit juices. The high level of choice of distractor C in Finland (36%), Sweden (31%), England (27%) and Australia (21%) suggests the possible widespread use of pull-ring openers for cool drink containers in these countries.

A set of chimes was made by cutting four pieces of pipe of different lengths from a long metal pipe and hanging them as shown in the picture below. Which of the pipes gave the **lowest** note when struck with a hammer?

A   Pipe X
B   Pipe Y
C   All gave the same note.
D   You cannot tell without trying.
E   It depends on where you hit it.

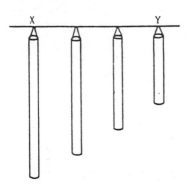

Country	N	A*	B	C	D	E	None
Australia	4259	56.97	20.85	1.72	6.85	11.71	1.90
Canada(Eng)	5104	58.24	17.34	2.65	9.95	10.57	1.25
England	3748	52.30	20.08	2.21	8.19	15.06	2.16
Finland	1600	54.02	20.53	1.42	6.18	17.18	0.67
Hong Kong	5342	39.44	32.24	6.31	14.74	5.89	1.38
Hungary	2590	47.37	14.36	3.32	17.16	16.04	1.75
Italy	5156	24.83	44.40	2.61	8.88	15.28	4.00
Japan	7924	53.16	32.59	2.08	5.37	5.97	0.82
Korea	3489	46.99	34.68	2.05	2.98	13.24	0.06
Norway	1305	58.63	15.46	1.46	6.49	14.20	3.75
Philippines	16851	30.34	44.54	6.47	11.05	6.93	0.67
Poland	4390	28.93	28.69	3.45	9.61	20.50	8.83
Singapore	5547	25.85	35.61	5.78	13.93	18.29	0.54
Sweden	1449	46.94	24.71	1.46	7.14	15.31	4.45
U.S.A.	2822	52.92	25.58	3.36	8.55	8.13	1.46

This item assesses knowledge and the correct response to the item is option A. Some of the high percentages choosing option B are of interest. Presumably, many 10-year old students associate small with low and large with high as on a numerical scale. This is clearly the common misunderstanding among children who have not had experience with chimes or pipes either inside or outside the classroom. However, the fact that option C was seldom chosen indicates that most children know that differing lengths of pipes produce different notes. A surprisingly high percentage of Polish students (9%) did not respond to this item.

POPULATION 2

Seven items have been selected for presentation. The first item is an earth science item and then there are two items each from biology, chemistry and physics. Whereas in Population 1 some children may not have had science as a formal school subject, Population 2 students did have science subjects at school.

The sun is the only body in our solar system that gives off large amounts of light and heat. Why can we see the moon ?

A	It is reflecting light from the Sun.
B	It is without an atmosphere.
C	It is a star.
D	It is the biggest object in the solar system.
E	It is nearer the Earth than the Sun.

Country	N	A*	B	C	D	E	None
Australia	4917	69.85	1.65	8.43	2.08	17.64	0.34
Canada(Eng)	5543	72.93	1.81	7.77	1.73	14.94	0.83
England	3118	53.47	2.85	10.48	3.82	28.60	0.78
Finland	2546	82.74	2.56	4.80	2.73	6.98	0.19
Hong Kong	4973	76.54	1.01	5.25	2.49	14.55	0.17
Hungary	2515	94.95	0.63	1.09	0.98	1.97	0.37
Italy	3228	62.22	3.50	10.56	1.80	18.57	3.35
Japan	7610	70.97	9.64	2.30	2.25	14.71	0.13
Korea	4522	62.95	3.43	13.93	2.41	17.21	0.07
Netherlands	5025	74.33	3.28	6.88	3.88	11.10	0.55
Norway	1420	68.12	2.37	5.01	2.45	20.67	1.38
Philippines	10874	71.66	1.86	3.85	10.15	12.28	0.19
Poland	4520	71.61	2.08	12.18	4.10	9.17	0.85
Singapore	4430	88.56	1.33	1.79	1.59	6.50	0.23
Sweden	1461	85.68	1.53	4.22	1.61	6.87	0.10
Thailand	3780	59.99	3.70	8.33	1.55	26.32	0.11
U.S.A.	2519	69.53	2.43	10.41	2.76	14.58	0.29

This item was also given to students in Population 1. It is a relatively easy item for Population 2 students but the increase in performance from Population 1 to Population 2 ranges from 4 percent in Korea and the U.S.A. to 33 percent in Hong Kong.

Moreover, the responses from the 17 countries would appear to show differences between those countries where the content associated with this aspect of the solar system, such as Hungary (95%) and Singapore (89%) is clearly taught to all students, and those countries where it is only taught to some students, for example England (53%), or where the information necessary to answer this question is gained from the media and reference books. It should be noted that option E remains a strong distractor in England (29%), Thailand (26%) and Norway (21%).

A girl wanted to learn which of three types of soil (clay, sand and loam) would be best for growing beans. She found three flower pots and tilled each with a different type of soil. She then planted the same number of beans in each, as shown in the drawing. She placed them side by side on a window sill and gave each pot the same amount of water.

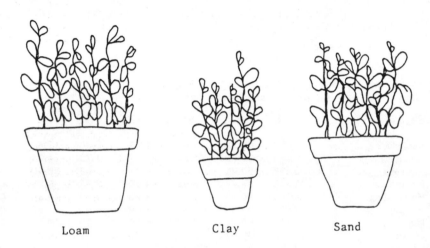

Loam   Clay   Sand

Why was the experiment <u>not</u> a good one for the purpose?

A   The plants in one pot got more sunlight than the plants in the other pots.
B   The amount of soil in each pot was not the same.
C   One pot should have been placed in the dark.
D   Different amounts of water should have been used.
E   The plants would get too hot on the window sill.

Country	N	A	B*	C	D	E	None
Australia	4917	8.81	65.24	5.17	14.06	5.79	0.94
Canada(Eng)	5543	7.31	72.30	5.25	11.68	2.52	0.95
England	3118	11.13	53.70	6.95	18.84	8.16	1.22
Finland	2546	9.51	54.10	3.64	26.65	5.39	0.71
Hong Kong	4973	8.70	61.55	5.34	20.04	4.17	0.20
Hungary	2515	3.13	62.27	2.87	26.95	3.69	1.08
Italy	3228	8.37	43.59	4.99	35.46	4.99	2.60
Japan	7610	12.27	63.01	11.63	9.62	3.20	0.26
Korea	4522	12.48	53.58	6.66	12.25	14.96	0.06

Country	N	A	B*	C	D	E	None
Netherlands	5025	9.63	64.61	3.66	17.49	3.87	0.74
Norway	1420	5.23	33.78	4.17	46.66	9.46	0.70
Philippines	10874	16.49	56.63	6.98	7.96	11.45	0.50
Poland	4520	9.46	54.05	2.90	25.23	6.56	1.81
Singapore	4430	13.37	62.60	4.62	12.21	6.52	0.69
Sweden	1461	7.15	52.65	9.78	24.38	4.72	1.33
Thailand	3780	8.02	77.06	6.65	4.39	3.77	0.11
U.S.A.	2519	10.42	62.00	7.73	15.00	4.21	0.65

This item is an interesting one from the perspective of science as an investigatory process, since it involves consideration of a controlled experiment. The correct answer is B and the item is relatively easy and with relatively homogeneous performances across countries. However, it seems likely that responses to this question from the different countries indicate the extent to which science, and in particular biology, is taught as a subject involving experimentation. The high performance of the Thai students (77%) suggests that this is a feature of their science courses, while the relatively low performance of the Italian (44%) and the Norwegian (34%) students would appear to indicate that such investigatory activities are not commonly included in the science courses of these countries at the lower secondary school level.

Which of the cells shown below would commonly be found in the human nervous system?

A   B   C   D   E

Country	N	A*	B	C	D	E	None
Australia	4917	24.48	11.04	15.12	21.53	25.12	2.71
Canada(Eng)	5543	43.69	10.58	8.52	16.52	18.21	2.48
England	3118	22.03	14.83	15.20	20.33	22.56	5.05
Finland	2546	12.17	9.03	17.13	27.97	32.54	1.15
Hong Kong	4973	8.38	22.99	20.59	21.02	26.39	0.64
Hungary	2515	77.51	3.84	4.15	6.20	7.42	0.88
Italy	3228	59.85	10.88	5.54	5.16	9.85	8.72

Country	N	A	B*	C	D	E	None
Japan	7610	47.52	13.28	15.74	8.81	14.33	0.32
Korea	4522	67.76	5.62	8.36	9.47	8.74	0.05
Netherlands	5025	36.43	10.40	7.53	19.88	22.64	3.12
Norway	1420	32.06	8.12	6.24	22.27	28.48	2.83
Philippines	10874	23.50	13.79	18.67	10.33	33.36	0.36
Poland	4520	69.17	3.94	4.77	2.12	18.23	1.76
Singapore	4430	18.81	19.87	15.92	23.95	20.81	0.63
Sweden	1461	21.12	17.86	9.82	23.70	24.59	2.91
Thailand	3780	10.90	22.14	17.41	16.87	32.47	0.21
U.S.A.	2519	27.04	16.12	15.69	18.98	21.12	1.05

The correct answer to this question is option A, and the item assesses knowledge of the shape and structure of cells. It is a relatively difficult biology item for 14-year old students except Hungary (78%), Poland (69%) and Korea (68%). In these three countries the structure of the cells of the human nervous system would appear to be included in the science curriculum at the lower secondary school level.

This topic had not been covered in the Hong Kong curriculum when the test was taken and, hence, the guessing as can be seen by the percent selecting each option.

Paint applied to an iron surface prevents the iron from rusting. Which <u>one</u> of the following provides the best reason?

A   It prevents nitrogen from coming in contact with the iron.
B   It reacts chemically with the iron.
C   It prevents carbon dioxide from coming in contact with the iron.
D   It makes the surface of the iron smoother.
E   It prevents oxygen and moisture from coming in contact with the iron.

Country	N	A	B	C	D	E*	None
Australia	4917	4.77	6.03	6.15	4.23	77.51	1.31
Canada(Eng)	5543	4.53	5.59	7.20	2.28	78.82	1.58
England	3118	4.17	6.75	7.63	4.43	75.39	1.63
Finland	2546	5.96	3.68	2.98	1.64	85.14	0.60
Hong Kong	4973	5.44	3.59	7.84	4.13	78.65	0.35
Hungary	2515	1.73	1.51	4.07	1.27	90.72	0.69
Italy	3228	5.94	4.41	6.79	1.99	77.55	3.32
Japan	7610	7.14	5.50	5.37	3.05	78.49	0.45
Korea	4522	3.17	3.06	5.09	3.29	85.29	0.09

Country	N	A	B*	C	· D	E	None
Netherlands	5025	3.56	2.31	5.11	1.48	86.95	0.59
Norway	1420	2.78	3.43	2.65	1.32	89.06	0.76
Philippines	10874	9.01	17.43	9.57	12.89	50.68	0.42
Poland	4520	4.67	2.53	5.30	3.28	82.45	1.77
Singapore	4430	4.16	5.53	11.28	4.55	73.81	0.68
Sweden	1461	5.02	5.22	5.52	1.63	81.81	0.80
Thailand	3780	7.06	9.37	8.78	2.84	71.86	0.09
U.S.A.	2519	7.25	8.77	11.13	5.61	66.27	0.97

This item was also given to 10-year old students in Population 1. The gains between Population 1 and Population 2 are quite high ranging from about 15 percent in the Philippines and Finland (although it should be noted that the performance in Finland was high in Population 1) to a 42 percent gain in Singapore. The low performance of the students in the Philippines (51%) should be noted, and it would appear that the topic of the rusting of iron is included in the science curriculum of only some of the students at the lower secondary school level in this country.

When 2 g (grams) of zinc and 1 g of sulphur are heated together, practically no zinc or sulphur remains after the compound zinc sulphide is formed. What happens if 2 g zinc are heated with 2 g of sulphur?

A   Zinc sulphide containing approximately twice as much sulphur is formed.
B   Approximately 1 g of sulphur will be left over.
C   Approximately 1 g of zinc will be left over.
D   Approximately 1 g of each will be left over.
E   No reaction will occur.

Country	N	A	B*	C	D	E	None
Australia	4917	29.83	28.83	7.11	12.09	18.94	3.19
Canada(Eng)	5543	22.83	31.56	7.18	11.68	20.58	6.18
England	3118	27.45	25.30	8.02	13.00	20.73	5.50
Finland	2546	30.53	31.45	8.78	11.01	17.01	1.22
Hong Kong	4973	29.58	21.91	12.59	12.73	22.07	1.11
Hungary	2515	25.90	42.01	4.51	13.59	12.57	1.42
Italy	3228	20.05	25.06	3.96	12.40	25.66	12.89
Japan	7610	21.31	55.56	7.14	6.57	8.88	0.55
Korea	4522	29.65	48.58	4.70	6.76	10.29	0.02
Netherlands	5025	18.72	43.77	8.95	8.21	17.17	3.18

Country	N	A	B*	C	D	E	None
Norway	1420	30.28	19.25	5.20	11.98	30.79	2.51
Philippines	10874	37.21	20.97	7.32	12.26	21.84	0.40
Poland	4520	14.52	38.60	5.80	17.44	18.29	5.35
Singapore	4430	26.80	33.72	6.48	6.19	25.81	1.01
Sweden	1461	25.06	23.25	6.75	6.84	34.18	3.92
Thailand	3780	32.57	32.79	4.11	9.10	21.30	0.13
U.S.A.	2519	22.65	28.78	10.16	16.24	21.29	0.88

This item is a relatively difficult one for fourteen year old students. The range of correct responses (B) is from 19 percent in Norway to 56 percent in Japan. Options A and E were the wrong answers most frequently chosen and both attracted significant numbers of students in most countries. It is clear that the basic ideas of chemical change have not been explained to all students at the lower secondary school level.

A set of chimes was made by cutting four pieces of pipe of different lengths from a long metal pipe and hanging them as shown in the picture below. Which of the pipes gave the lowest note when struck with a hammer?

A   Pipe X
B   Pipe Y
C   All gave the same note.
D   You cannot tell without trying.
E   It depends on where you hit it.

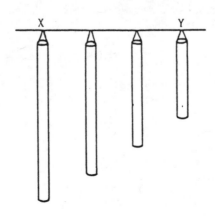

Country	N	A*	B	C	D	E	None
Australia	4917	73.82	15.24	1.11	3.23	6.32	0.28
Canada(Eng)	5543	71.85	13.23	1.76	4.40	7.74	1.03
England	3118	75.23	14.21	0.82	2.56	6.68	0.51
Finland	2546	59.39	22.07	1.44	5.70	11.12	0.27
Hong Kong	4973	56.60	24.86	2.08	7.31	8.62	0.53
Hungary	2515	71.27	12.56	0.95	8.92	5.47	0.83
Italy	3228	45.25	31.08	2.75	6.87	11.20	2.85
Japan	7610	65.28	24.76	0.94	3.23	5.45	0.33
Korea	4522	51.26	24.36	2.73	1.97	19.54	0.14
Netherlands	5025	76.39	15.41	1.47	2.02	4.02	0.68
Norway	1420	70.41	16.34	0.88	4.75	7.01	0.60
Philippines	10874	30.35	36.33	4.56	15.09	13.36	0.31
Poland	4520	43.49	32.15	3.00	9.68	9.43	2.25
Singapore	4430	51.77	25.04	2.31	6.22	14.39	0.27
Sweden	1461	59.19	28.93	1.59	2.77	5.71	1.80
Thailand	3780	38.82	26.19	5.60	14.97	14.37	0.05
U.S.A.	2519	65.11	17.86	3.72	6.63	5.48	1.21

This item was also given at the Population 1 level. For the fourteen year olds it is a relatively easy item with correct responses being between 30 percent in the Philippines and 75 percent in England. The increase from Population 1 to Population 2 ranged from 0 percent in the Philippines to 26 percent in Singapore. The most frequent wrong answer was the same as for Population 1, namely option B, indicating that the common misunderstanding that the shortest pipe produces the lowest note remains even after the content has been more extensively taught.

The objects P, Q and R of weight 15 N (newtons), 20 N and 7 N, are hung with a light thread as shown in the figure.

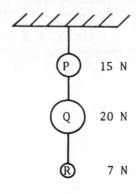

What is the tension in the thread between P and Q?

A	42 N
B	35 N
C	27 N
D	15 N
E	7 N

Country	N	A	B	C*	D	E	None
Australia	4917	10.18	42.44	28.27	7.75	7.49	3.88
Canada(Eng)	5543	10.84	41.58	29.87	7.30	6.18	4.23
England	3118	12.28	41.26	26.07	8.73	7.94	3.73
Finland	2546	11.09	44.38	26.41	8.50	7.93	1.69
Hong Kong	4973	11.94	42.69	28.21	8.76	7.50	0.91
Hungary	2515	13.28	25.68	52.01	3.37	4.16	1.51
Italy	3228	16.21	34.96	19.95	3.28	16.40	9.19
Japan	7610	20.64	22.36	51.24	3.34	2.08	0.35
Korea	4522	14.04	35.82	27.98	11.54	10.49	0.11
Netherlands	5025	21.28	10.21	45.04	12.61	7.28	3.57
Norway	1420	9.69	33.58	31.87	12.04	8.22	4.59
Philippines	10874	13.93	59.93	10.89	7.34	7.56	0.34
Poland	4520	18.49	34.09	33.25	4.97	5.03	4.17
Singapore	4430	10.59	51.48	21.95	7.89	7.19	0.89
Sweden	1461	7.18	39.21	27.72	14.24	7.30	4.34
Thailand	3780	17.57	53.79	17.54	5.99	5.09	0.02
U.S.A.	2519	9.74	46.85	19.59	12.15	10.34	1.34

The correct response to this item is alternative C. The item assesses the higher order cognitive skills of application and analysis. It is clear that this is a relatively difficult item for 14-year old students with the percentage correct ranging from 11 percent in the Philippines to 52 percent in Hungary. B is a strong distractor presumably because the students have not learned the principles involved and, because the question refers to P and Q, they just added 15 N and 20 N. It would appear that the analyses of gravitational forces acting in a sample system is only taught at this grade level in Hungary (52%), Japan (51%), the Netherlands (45%), Poland (33%) and Norway (32%). This item was also given to students at Population 3.

Population 3

The figures presented for the first three items are based on the results of the combined groups of biology, chemistry and physics students at Population 3 level.

When 2 g (grams) of zinc and I g of sulphur are heated together, practically no zinc or sulphur remains after the compound zinc sulphide is formed. What happens if 2 g zinc are heated with 2 g of sulphur?

A Zinc sulphide containing approximately twice as much sulphur is formed.
B Approximately 1 g of sulphur will be left over.
C Approximately 1 g of zinc will be left over.
D Approximately 1 g of each will be left over.
E No reaction will occur.

Country	N	A	B*	C	D	E	None
Australia	3881	17.96	69.23	3.79	3.35	4.33	1.34
Canada(Eng)	8943	14.56	70.74	4.43	3.91	4.82	1.54
England	2693	8.73	85.37	2.93	0.97	1.11	0.89
Finland	2552	21.24	60.82	5.41	4.70	6.19	1.65
Hong Kong (Form 6)	6025	10.06	81.66	5.39	1.31	1.05	0.53
Hong Kong (Form 7)	3679	9.43	84.70	3.86	0.71	0.92	0.38
Hungary	842	15.20	75.53	1.54	0.95	5.94	0.83
Italy	2130	13.90	56.10	3.00	2.68	13.94	10.38
Japan	3867	5.33	88.03	2.53	1.50	2.56	0.05
Norway	1002	20.36	69.26	2.79	2.69	3.39	1.50
Poland	3245	7.24	79.69	5.33	2.80	3.48	1.45
Singapore	2918	10.11	83.62	2.78	0.55	2.43	0.51
Sweden	2430	11.56	80.58	3.09	0.82	3.21	0.74

This item assesses understanding of the concept of chemical change and the correct answer is option B. The scores on this item range from 56 percent in Italy to 88 percent in Japan. This was one of the items reported earlier for Population 2. The lowest gain from Population 2 to Population 3 was 31 percent in Italy and the highest gain was 60 percent in England. It is surprising that there are so many science students in some countries who do not understand this important concept. Perhaps, however, it is desirable to have carried out laboratory experiments involving the heating of such substances together in order to understand what happens.

This item was answered by biology, chemistry and physics students at Population 3.

Which of the cells shown below would commonly be found in the human nervous system?

| A | B | C | D | E |

Country	N	A*	B	C	D	E	None
Australia	3881	57.85	6.11	3.22	18.09	12.55	2.19
Canada(Eng)	8943	71.81	4.32	2.23	10.15	9.38	2.11
England	2693	78.28	4.16	0.97	8.91	5.38	2.30
Finland	2552	80.25	0.94	0.98	9.01	8.46	0.35
Hong Kong (Form 6)	6025	88.38	2.17	2.34	5.21	1.68	0.22
Hong Kong (Form 7)	3679	86.25	2.47	2.94	7.07	1.06	0.22
Hungary	842	95.49	0.95	0.83	0.95	1.43	0.36
Italy	2130	77.04	4.04	2.77	2.91	5.26	7.98
Japan	3867	83.01	2.22	2.46	6.34	5.90	0.08
Norway	1002	60.18	4.39	0.50	16.87	16.37	1.70
Poland	3245	95.01	0.59	0.62	0.80	2.74	0.25
Singapore	2918	71.14	7.13	3.56	11.38	6.00	0.79
Sweden	2430	51.28	5.51	1.23	23.95	17.00	1.03

The correct response to this item is option A, and the question is the same as that given to students at the Population 2 level.

This item was an easy one for Population 3 with scores ranging from 51 percent in Sweden to 95 percent in Hungary. Hungarian students in Population 2 had, however, performed well on this item (78%), and therefore, Hungary had the smallest gain of 17 percent, as did Italy, from Population 2 to 3. The highest gain was 80 percent in Hong Kong (Form 6).

This item was answered by biology, chemistry and physics students at Population 3.

The objects P, Q and R of weight 14 N (newtons), 20 N and 7 N, are hung with a light thread as shown in the figure.

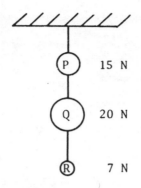

What is the tension in the thread between P and Q?

A   42 N
B   35 N
C   27 N
D   15 N
E   7 N

Country	N	A	B	C*	D	E	None
Australia	3881	12.39	22.21	56.58	3.45	2.40	2.96
Canada(Eng)	8943	11.42	22.40	56.46	3.76	2.55	3.42
England	2693	10.55	9.58	75.64	1.75	1.00	1.49
Finland	2552	11.72	21.87	55.41	4.74	2.90	3.37
Hong Kong (Form 6)	6025	5.21	3.04	87.54	3.05	0.90	0.27
Hong Kong (Form 7)	3679	5.30	2.04	89.54	2.39	0.60	0.14
Hungary	842	6.29	9.03	80.17	2.38	1.43	0.71
Italy	2130	17.04	11.60	58.69	1.03	6.24	5.40
Japan	3867	11.87	9.36	73.11	3.18	1.99	0.49
Norway	1002	10.38	14.47	61.88	4.19	2.40	6.69
Poland	3245	9.68	13.96	71.37	2.71	1.14	1.14
Singapore	2918	14.12	11.62	68.71	3.36	1.44	0.75
Sweden	2430	7.33	6.67	79.05	3.25	1.56	2.14

To answer this question correctly by marking option C, requires the ability to apply knowledge and understanding of the simple principles of statics which are taught in physics courses. The lowest and highest scores were in Finland (55 %) and Hong Kong (90%). The gains from Population 2 to 3 levels ranged from 22 percent in Japan to 61 percent in Hong Kong.

This item was answered by biology students only at Population 3.

What initially determines whether a human baby is going to be a male or a female?

A   The DNA in the sperm.
B   The DNA in the egg.
C   The RNA in the sperm.
D   The RNA in the egg.
E   The DNA and RNA in both sperm and egg.

Country	N	A*	B	C	D	E	None
Australia	1631	39.06	6.37	6.50	1.49	44.90	1.69
Canada(Eng)	3254	44.58	5.42	6.42	1.07	41.92	0.59
England	884	47.66	4.86	3.53	0.28	43.18	0.49
Finland	1652	58.06	4.31	10.21	1.56	25.20	0.66
Hong Kong (Form 6)	5960	43.46	5.84	13.10	2.68	33.73	1.19
Hong Kong (Form 7)	3614	50.93	4.65	11.15	2.28	29.58	1.39
Hungary	301	87.08	1.71	0.78	2.15	7.42	0.86
Italy	147	33.35	3.71	2.43	1.16	51.08	8.27
Japan	1212	37.95	12.21	8.58	3.05	37.95	0.25
Norway	276	46.65	3.44	9.06	0.82	39.15	0.89
Poland	764	73.96	5.36	5.40	0.61	14.32	0.36
Singapore	902	62.64	4.10	2.77	0.89	29.05	0.55
Sweden	1232	34.63	5.57	13.57	2.57	41.70	1.96
U.S.A.	659	48.44	6.00	9.17	2.72	33.25	0.42

The correct response to this item is option A. To answer the item correctly requires knowledge of key biological ideas. These ideas were well known in Hungary (87%) and Poland (74%) and not well known to biology students in Italy (33%) and Sweden (35%). Only in Hungary and Poland did distractor E fail to attract very large numbers of students. These results should be of interest to curriculum developers and textbook authors at this level.

This item was answered by biology students only at Population 3.

In order to obtain two crops in one growing season a farmer planted some seeds which he had harvested the previous week but the seeds failed to germinate. What can be concluded from this observation?

A   The farmer did not provide the right conditions for germination.
B   The seeds needed a longer period of maturation.
C   The farmer had not removed inhibiting substances.
D   The seeds required a period of low temperature.
E   The data are inadequate for a conclusion to be reached.

Country	N	A	B	C	D	E*	None
Australia	1631	16.28	26.05	3.09	2.77	50.33	1.48
Canada(Eng)	3254	10.16	30.10	3.43	4.28	50.16	1.87
England	884	9.56	22.76	2.56	10.87	54.07	0.19
Finland	1652	9.18	32.18	4.18	31.76	21.69	1.01
Hong Kong (Form 6)	5960	20.83	19.24	14.26	5.08	38.59	1.99
Hong Kong (Form 7)	3614	18.55	18.18	11.30	7.80	41.80	2.38
Hungary	301	13.04	49.62	11.04	1.99	22.56	1.75
Italy	147	13.09	38.20	1.95	16.33	27.15	3.28
Japan	1212	10.73	33.09	3.96	20.13	31.93	0.17
Norway	276	5.12	57.55	2.59	3.50	28.42	2.83
Poland	764	6.72	43.28	3.40	13.55	32.10	0.96
Singapore	902	8.98	18.96	4.32	1.44	65.96	0.33
Sweden	1232	5.31	49.72	4.68	6.49	32.21	1.58
U.S.A.	659	14.02	24.13	4.28	3.39	53.39	0.79

The correct response to this item is option E, and to answer the question a student is required to analyze the alternatives and to recognize that insufficient evidence is provided. It is of interest to note the low level of performance in Finland (22%) and Hungary (23%) and the strong attraction of option D (32%) in Finland and option B in both Hungary (50%) and Finland (32%). Option B is the most common distractor in most countries, and this common misunderstanding warrants further consideration by curriculum developers. The high score in Singapore (66%) is of interest.

This item was answered by chemistry students only at Population 3.

The graph shows the solubility of two substances X and Y. A sample of 150 g of X and 75 g of Y is placed in a beaker containing 100 cm3 of water. Assume that the placing of the two substances together has no effect on how either dissolves. The mixture is filtered at 60oC. What would the residue on the filter paper consist of?

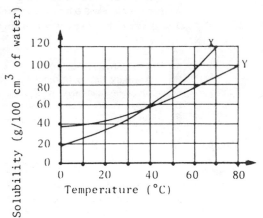

A   95 g of X and 15 g of Y
B   55 g of X and 75 g of Y
C   95 g of X
D   75 g of Y
E   55 g of X

Country	N	A	B	C	D	E*	None
Australia	1177	14.59	12.82	11.45	6.38	49.84	4.92
Canada(Eng)	2923	16.47	14.12	13.18	6.68	43.57	5.97
England	892	8.36	9.90	10.51	4.22	64.33	2.68
Finland	971	24.17	15.49	17.41	7.48	30.10	5.35
Hong Kong (Form 6)	6018	8.22	7.87	6.28	4.00	71.56	2.07
Hong Kong (Form 7)	3670	4.54	4.52	3.35	2.19	82.55	2.85
Hungary	143	13.46	15.59	4.81	7.19	55.83	3.13
Italy	217	14.27	12.68	7.72	1.86	37.73	25.74
Japan	1468	4.84	6.40	6.74	4.43	76.98	0.61
Norway	283	20.15	10.36	7.07	5.93	49.29	7.20
Poland	765	20.69	14.65	12.26	8.19	40.06	4.14
Singapore	945	9.63	6.98	4.87	2.43	74.92	1.16
Sweden	1172	8.93	10.40	10.39	5.64	60.77	3.87
U.S.A.	537	13.96	14.91	18.39	10.56	34.78	7.40

This question requires the skill of interpretation of the graphical information provided, together with an understanding of the ideas of solubility. The high level of performance in Hong Kong (F6: 72%, F7: 83%), Japan (77%) and Singapore (75%) is of interest. The lower levels of correct response by students particularly in Finland (30%) and the United States (35%) as well as other countries suggests lack of experience with such graphical material. This view is confirmed by a high level of non-response, in Italy (26%).

This item was answered by chemistry students at Population 3.

A compound X has the formula C_3H_8O. On partial oxidation it changes to C_3H_6O. From this information, which of the following is the most likely description of X?

A   An aldehyde (alkanal)
B   A tertiary alcohol (alkanol)
C   An olefin (alkene)
D   A secondary alcohol (alkanol)
E   An ether

Country	N	A	B	C	D*	E	None
Australia	1177	19.27	14.43	10.41	36.43	10.89	8.58
Canada(Eng)	2923	16.44	18.39	15.63	19.07	20.77	9.70
England	892	15.74	6.61	1.62	67.43	7.98	0.61
Finland	971	23.42	15.64	17.36	21.28	19.12	3.19

Country	N	A	B	C	D	E*	None
Hong Kong (Form 6)	6018	17.77	23.05	5.92	39.29	11.89	2.07
Hong Kong (Form 7)	3670	10.48	6.42	1.83	71.45	6.92	2.90
Hungary	143	20.95	21.02	12.78	23.84	14.05	7.36
Italy	217	22.04	9.24	7.96	38.71	8.39	13.65
Japan	1468	17.78	14.37	6.81	37.67	22.21	1.16
Norway	283	24.78	6.81	12.86	44.22	8.44	2.90
Poland	765	20.23	11.17	10.64	45.35	8.05	4.56
Singapore	945	20.21	8.25	3.17	59.15	8.25	0.95
Sweden	1172	27.04	14.91	10.80	20.61	19.85	6.79

The correct response to this item is option D. The question requires a knowledge of the principles of organic chemistry and the ability to apply that knowledge. The increase in level of achievement between Form 6 (39%) and Form 7 (71%) in Hong Kong and the high levels of achievement in both England (67%) and Singapore (59%) would seem to indicate that organic chemistry is an important component of chemistry courses in these countries and of lesser importance in other countries. The results for Finland (21%), Hungary (24%), Canada (English) (19%) and Sweden (21%) would seem to indicate that organic chemistry is not taught in these chemistry programs.

This item was answered by physics students at Population 3.

A stone is dropped from rest down a deep well. It takes 2 s to reach the bottom. How deep is the well?.

Assume that the air resistance on the falling stone is negligible and that the accelleration due to gravity g = 9.8 m.s-2

A   4.9 m
B   9.8 m
C   19.6 m
D   39.2 m
E   78.4 m

Country	N	A	B	C*	D	E	None
Australia	1073	2.07	3.61	88.11	5.40	0.11	0.71
Canada(Eng)	2766	5.91	5.70	74.33	13.05	0.26	0.76
England	917	0.83	2.53	90.50	5.45	0.49	0.21
Finland	810	7.30	9.06	63.42	18.59	0.55	1.08

Country	N	A	B	C	D	E*	None
Hong Kong (Form 6)	6025	0.99	2.97	91.95	3.36	0.57	0.15
Hong Kong (Form 7)	3679	0.59	2.49	92.68	3.31	0.75	0.19
Hungary	398	2.29	3.15	88.24	5.63	0.00	0.70
Italy	1766	11.91	6.00	65.82	11.30	0.77	4.20
Japan	1187	0.67	1.94	86.18	10.87	0.34	0.00
Norway	443	1.30	4.30	82.93	9.54	0.80	1.13
Poland.	1716	5.18	4.02	77.78	11.74	0.94	0.34
Singapore	1071	0.75	2.33	91.78	4.58	0.19	0.37
Sweden	1156	5.87	7.83	53.42	30.74	0.80	1.35
U.S.A.	485	4.57	4.50	74.06	15.91	0.40	0.57

This question involves the application of the laws of kinematics to a very simple problem in physics. The correct answer is option C. In general, the question was an easy one for the physics students in most countries. The low level of correct response in Sweden (53%) and the popularity of option D (31%) should be of concern to the authors of textbooks in that country.

This item was answered by physics students at Population 3.

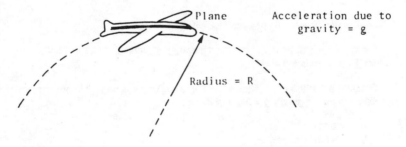

An aircraft flies in a vertical circular path of radius R at a constant speed. When the aircraft is at the top of the circular path the passengers feel "weightless". What is the speed of the aircraft?

A   gR

B   √gR

C   g/R

D   √g/R

E   2gR

Country	N	A	B*	C	D	E	None
Australia	1073	16.84	39.91	9.45	14.50	10.71	8.58
Canada(Eng)	2766	17.58	28.09	14.62	13.71	15.22	10.77
England	917	16.88	47.63	7.31	15.13	8.87	4.18
Finland	810	20.35	29.60	9.58	13.93	19.14	7.40
Hong Kong (Form 6)	6025	4.13	72.48	3.85	13.06	5.36	1.12
Hong Kong (Form 7)	3679	1.81	85.34	1.65	7.61	2.82	0.78
Hungary	398	1.72	62.66	11.36	16.89	3.91	3.46
Italy	1766	30.87	13.94	10.43	5.72	12.25	26.79
Japan	1187	12.81	49.28	6.66	19.46	9.52	2.27
Norway	443	13.33	54.22	5.32	9.42	12.89	4.82
Poland	1716	11.25	57.43	5.69	10.74	10.79	4.10
Singapore	1071	8.50	46.78	4.58	25.68	9.24	5.23
Sweden	1156	29.16	26.11	8.68	11.80	15.85	8.40
U.S.A.	485	14.12	35.50	7.83	13.56	14.11	14.87

The correct response to this question is option B. The item requires an understanding of the principles of circular motion in physics and proved surprisingly difficult for students in all countries except Hong Kong (Form 6: 72%; Form 7: 85%) and Hungary (63%). Only in Italy (14%), however, would it appear that students tested had not been exposed to the ideas involved.